FALKLAND ISLANDS AND PATAGONIA
JASON ISLANDS
WEST FALKLAND ISLAND
EAST FALKLAND ISLAND
FALKLAND SOUND
KING GEORGE BAY
QN. CHARLOTTE BAY
WEDDELL I.
SAUNDERS I.
PEBBLE IS.
Byron Sound
Port North
Westpoint I.
Keppel Sd.
C. Dolphin
Foul B.
P. San Carlos
Grantham Sound
Brenton Loch
Wickham Heights
BERKELEY SOUND
PORT SALVADOR
Stanley Har.
Port Fitz Roy
Port Harriet
Port Pleasant
CHOISEUL SOUND
Useless B.
Lively Sound
ADVENTURE SOUND
BAY OF HARBOURS
Port Albemarle
Port Stephens
Speedwell I.
George I.
Barren I.
Sea Lion Is.
Bleaker I.
Hornby Hills
P. Philomel
P. Purvis
Carcass I.

FROM THE FALKLANDS TO PATAGONIA

This book is dedicated to the memory of
Mabel Halliday de Miller and Enrique

Note on sources and quotations used

Unless otherwise attributed, all prose quotations in
this book are from the diaries, documents or
comments of Mabel, other members of the Halliday
family and neighbours. Verse quotations are taken
from Jose Hernández's *El Gaucho Martín Fierro*,
Argentina's famous epic of life on the pampas. The
sources of other quotations are listed in the
Bibliography.

From The Falklands To Patagonia

THE STORY OF A PIONEER FAMILY

by

MICHAEL JAMES MAINWARING

Allison & Busby

London / New York

First published 1983 by
Allison and Busby Limited,
6a Noel Street, London W1V 3RB, England,
and distributed in the USA by
Schocken Books Inc.,
200 Madison Avenue, New York, N.Y. 10016

This book is published
with the aid of a subvention from the
Latin American Publications Fund

British Library Cataloguing in Publication Data
Mainwaring, Michael
    From the Falklands to Patagonia.
    1. Falkland Islands — Emigration and immigration
    2. Patagonia — Emigration and immigration
    I. Title
    997'.11   F3031
    ISBN 0-85031-488-7

Set in Goudy Old Style by Top Type Phototypesetting Co. Ltd., London
and bound and printed in Great Britain by
Richard Clay (The Chaucer Press) Ltd., Bungay, Suffolk.

Contents

Page

Illustrations

Unless otherwise stated, the photos used are from collections belonging to the Halliday, Rudd, Jamieson and Felton families of southern Patagonia.

Preface

I FIRST met Mabel Halliday de Miller at the sheepfarm Hill Station, near the town of Río Gallegos in the province of Santa Cruz, Argentina, in 1966. I was on vacation in Patagonia from St George's College, Buenos Aires, with a colleague whose sister had married Jimmy Halliday, Mabel's nephew. "Aunty Mabel" was the last surviving child of William Halliday, who had emigrated from Dumfries in Scotland to the Falklands as a shepherd in 1862, and from there to Patagonia in 1885. Still living in the wooden house in which she had been born in 1888, Mabel told me the story of her family's arrival and the establishment of the sheepfarm. She also loaned me various diaries, documents and photographs. As I was already interested in the area, I decided to research its history more fully in order to understand the details and significance of her family's achievements. I visited desendants of original settlers around Río Gallegos and in various parts of southern Patagonia. I also carried out research in Chubut, Buenos Aires, Oxford, Dumfries and London.

I should like to acknowledge my gratitude to Martin Garvie for introducing me to Patagonia, to Jimmy and Christine Halliday for their hospitality at Hill Station, to Mabel's sister-in-law Helen Halliday, to the Rudd family at Cape Fairweather, to Bill and Yolanda Jamieson at Moy Aike, to the Tweedies at Stag River, to Tegai Roberts, curator of the Welsh Museum in Gaiman, and to Gordon and the late Catherine Garsed, Archie and Amalia Norman and to Helen and the late Peter Street in Buenos Aires.

I am grateful to Radio Río Gallegos, the National Library of Buenos Aires, the *Buenos Aires Herald*, Canning House in London, the Welsh Argentine Society in Britain, the Dumfries and Galloway Regional Council, and the Bodleian Library and the Latin American Centre in Oxford.

I would also like to thank: Michael Arbon, Simon Astaire, Nick Bruce, Simon Burridge, Jane Cameron, Professor Raymond Carr, Malcolm Deas, Edward Greene, Doctor Martin Henig, Axel Klein, James McNeilly, Julian Munby, Ann and Helen Nimmo-Smith, Peter Parkes, Francis Prieur, Olliver Robinson, Mark Sedgwick, David Stevens and Jonathan Wright.

I am especially grateful to Jerry Moeran and Jean Hunt for reproducing the maps and photographs, and my parents and family for their continual support.

M.J.M.
June 1983

1

Prologue: A Patagonian Welcome

"If you go to sea by your own sweet will,
It's silly to fear a wetting." (Hernández)

ON THE morning of 31 July 1885 a small steamship sidled towards an estuary in the south of the south of South America. Capt. Winther of the *Ranée* was paying proper respect to a coastline infamous for shipwreck and disaster. And he had special cause for watching anxiously as the vessel crossed the mile-wide bar between Loyola Point and Cape Fairweather: he was carrying on board a family who had taken the unprecedented decision to settle on the banks of the estuary of the Río Gallegos.

William Halliday, his wife Mary, their seven young children and Mary's father William McCall had brought with them from their previous home in the Falkland Islands many boxes of provisions and six favourite sheepdogs. William himself had visited the area on an exploratory trip the previous summer, but for the rest of the family this was their first glimpse of what had been almost the only subject of conversation for the last few months: Patagonia. And it was not an encouraging introduction. Although discovered over 350 years earlier during Magellan's circumnavigation of the globe, the Río Gallegos's only sign of habitation was a corrugated-iron shack a few hundred yards up from the south beach, refuge for stranded sailors. There was not a tree in sight. Snow stretched remorselessly towards the west and south. But it was to the hills on the north bank that the passengers looked most keenly, for it was there that William had recently leased land from the Argentine government and now proposed to establish a sheepfarm, which he had already named Hill Station.

The *Ranée* dropped anchor in the sanctuary of the harbour. Her two tenders were lowered to carry the settlers and their possessions to the north bank, less than a mile away. William, two of the sheepdogs and four oarsmen from the crew were in the first boat, "Grandpa" McCall and the eldest son, eight-year-old Willie, were in the second which was piled high with as many provisions as seemed safe: tents, blankets, tools, a sack of sugar, another of salt, boxes of ammunition and candles, and a few schoolbooks. The second boat would be towed across by the first. The weather was bitter but bright, with no sign of the notorious Patagonian winds. Huddled together, the rest of the family watched with Capt. Winther from the deck of the *Ranée*. In all the excitement it had been temporarily forgotten that today was Agnes Jane's tenth birthday.

The two boats set out, but less than halfway across the estuary they were struck by a storm that came out of nowhere. The second boat began to flounder and, with William frantically hauling her in, Grandpa McCall and Willie scrambled from one to the other. They then had no option but to cut the towline and watch the boat with its precious cargo disappear under the waves. The weather just as suddenly was calm again. Now precariously overloaded, the surviving boat reached the safety of the north bank.

During the afternoon, the rest of the family, the dogs and possessions were rowed across. There was no further incident that day.

But Patagonia's characteristic welcome had only just begun.

With the tents and blankets lost, the Hallidays established their first home among a clump of barberry bushes in a small gully a few hundred yards up from the beach. Capt. Winther presented them with a sack of ship's biscuit and with a sail for added protection against the cold, but that evening, being under contract to return to the Falklands as soon as possible, he was obliged to take advantage of the rising tide and to leave the settlers to their fate. He steamed out of the estuary under a full moon.

The Hallidays knew that they would have to face many hazards in Patagonia, but they were unaware that the spring-tides of the Río Gallegos are among the highest in the world, rising to over 50 feet above the ebb. They simply stacked the remaining possessions high up on the beach out of harm's way, as they thought, and fell asleep under their improvised tent. During the night the tide advanced.

On their first Patagonian dawn the Hallidays woke to discover that out of all the provisions brought so painstakingly from the Falklands there remained a large kettle, a Singer sewing-machine and Grandpa

McCall's box of tools. The rest had been swept away.

William was not a man who despaired easily but on that morning of 1 August 1885 he must have shuddered at his predicament. He had intended to kill some of the abundant wildlife but the rifles and ammunition had disappeared in what the family came to refer to simply as "that tragedy". The only food he had with which to feed his seven children, of whom the youngest, Archie, was nine months old, was the ship's biscuit. There would be no outside help. An Indian encampment was said to exist somewhere to the north and its inhabitants were reputedly friendly, but they were still the devils William did not know and did not yet wish to know. Otherwise, the nearest "neighbours" were a German who had pitched his tents to the south of the Río Gallegos, over forty miles away by the nearest crossing-point, and some Spaniards who had settled sixty miles to the north. Even if these people could be reached, there was no guarantee that they would be in a position to help. The closest centres of civilization were the Chilean town of Punta Arenas on the Straits of Magellan 120 miles to the southwest, and the recently-established seat of the governor of the Argentine territory of Santa Cruz at Los Misioneros some 200 miles to the north on the Atlantic coast. It was highly unlikely that any ship would visit the Río Gallegos for the next few months.

Above all, the sheepfarm of the Hallidays' dreams did not as yet have a single sheep. Rather than transport the animals direct from the Falklands, William had arranged to travel south to the Straits of Magellan to purchase a flock from a farmer already well-established there. But now even that plan was in jeopardy. William was reluctant to leave his young family to the vagaries of a Patagonian winter without adequate food or shelter. He must have questioned, and even regretted, his decision to leave the security of a job in a British colony for the sake of settling in what had persistently proved to be one of the least hospitable parts of the world. Patagonia! Almost half a million square miles of wilderness, unexplored, unmapped and uninhabited except by Indians, by a handful of evangelists, by groups of settlers at its northern and southern extremities, by pumas, by huge birds, by strange beasts, even by prehistoric creatures, and, if the rumours of three centuries contained a seed of truth, by a race of human giants. Patagonia! On the one hand vast enough to accommodate man's wildest dreams and ambitions, on the other hand synonymous with foul weather, desolation and death.

William had chosen to settle in such a place, after due deliberation, because of its rich pastures and potential. Now he and his family were likely to starve. Why had he done it?

THE STORY starts in Scotland. William Halliday was born on 10 October 1845 in the parish of Johnstone, Dumfriesshire. Only son of John Halliday, a sailor, and Margaret née Gillespie, he was taught from an early age to be proud of his Border heritage: his ancestors had adopted the war-cry "A Holy Day!" because a day spent in rout and slaughter on the English border was considered a truly holy one, and when Richard the Lionheart sent a thousand men from nearby Annandale to fight in the Crusades "near all of them were Hallidays". With his father often away at sea, William lived with his mother and grandparents at 47 English Street, Dumfries, helping to run the Gillespie cab and stage-coach business. Whether from the constant comings and goings of the passengers, from the stories of exotic places told by his father on rare visits home, or from the sight of so many of his contemporaries leaving Scotland for far corners of the globe, William soon acquired his own hankering for travel. His heart was set on India but towards the end of 1861, when he was sixteen years old, he overheard two customers talking about the Falkland Islands, far away in the South Atlantic. He scanned the atlas, made enquiries and learned that the Falkland Islands Company, or FIC, was offering free passage and accommodation, a reasonable wage and a pension after twenty years' work, to young men willing to emigrate to the islands as shepherds.

Twenty years was a long time to sign away and, although an adequate rider, William knew next to nothing about sheep. His mother was upset at the idea as it had been assumed that the only son would eventually take over the family business. But the Falkland Islands appealed to William, perhaps because he thought he would enjoy more success in such a remote spot, perhaps because of the very remoteness itself. Besides, what else was he to do? He had no intention of spending the rest of his days in a provincial city organizing other people's travel. "If the Hallidays of Annandale could march in their hundreds to the Crusades...."

His mind was made up. In April 1862 he travelled to Liverpool, signed a contract at the offices of FIC and, by the end of the month, was a passenger on RMS *Boyne* bound for South America. He dutifully wrote home to his parents from each port on the three-month journey. In Rio de Janeiro the *Boyne* was loaded with coal by black slaves: "The slave-driver had a wee white dog which bit the heels of the slaves as the driver hit them with his whip. One poor fellow fell over the pier and was left there hanging to die." At Santos the yellow-fever epidemic had recently been so severe that many of the ships lay without crews: "More crews arrived, but they died too." Eventually the Brazilian authorities had taken all the ships outside the port and sunk them.

The few other contract shepherds on board considered William rather serious as he touched neither alcohol nor tobacco, and, unlike the majority, knew how to write. They also noticed that he used the telescope with some difficulty with his left eye: "In mid-Atlantic they placed a hair across the lens and told me it was the line of the Equator." William seems to have been the sort of young man who would have quietly led the others to believe that he had believed them.

In Montevideo the few passengers heading for the Falklands were transferred to the sailing-ship *Black Hawk*: "She was built as a slave-runner, with ring-bolts still in the deck. She could sail over the waves like a gull and was quick to get away if chased by gunboats." At last, towards the end of July 1862, in the depths of the southern winter, William arrived at what was likely to be his home for the next twenty years.

The Falklands, consisting of two main islands and almost 350 islets, were shrouded in historical uncertainty and controversy appropriate to their remoteness. They had been first sighted perhaps by the Florentine adventurer Amerigo Vespucci as early as 1504 or, more likely, by the English explorer John Davis in 1592 when blown off course in his ship the *Desire*. They were first known as Davis's Southern Islands or as Hawkins's Maidenland, after Sir Richard Hawkins had dedicated them to the Virgin Queen Elizabeth in 1593, or as the Sebaldines, after the Dutch navigator Sebald van Weerdt. In the late seventeenth century they became known to the English as Falkland's Islands, and subsequently as the Falkland Islands or the

Falklands, after Captain John Strong passed between the two main islands in 1690 and named the passage Falkland Sound in honour of Anthony Cary, 5th Viscount Falkland, Treasurer of the Navy at the time. In the eighteenth century they were also named the Malouines after visits by Frenchmen from St Malo. The Spanish adapted this name to Las Malvinas, calling East Falkland Isla Soledad and West Falkland Isla Gran Malvina.

Claims to sovereignty had been as diverse as the names. The French established a settlement at Port Louis on East Falkland in 1764. The British settled on West Falkland in 1766, in which year France yielded its rights to Spain, who in 1770 expelled the British. The following year the British colony was reinstated, but three years later they withdrew again, leaving a lead plaque on West Falkland claiming sovereignty. Spain controlled both islands until 1811 when her rule in what is now Argentina ended and the United Provinces of the Río de la Plata were established. In 1820 the government of Buenos Aires sent a ship and claimed the then unoccupied islands. They established a settlement on East Falkland, appointed a ''Commandant of the Malvinas'' and claimed sovereignty over both islands, a claim in 1829 disputed by Britain. In 1831 the Argentines seized three American sealing-vessels and, in retaliation, the American warship *Lexington* destroyed the settlement. In December 1832 Britain reclaimed West Falkland and a month later was in control of East Falkland.

Whatever the discrepancies of the history of the islands and the rights of ownership, the fact was that the British had controlled the Falklands since early 1833 and had gazetted them as a colony in 1841. As far as William Halliday and any other shepherds arriving in the 1860s were concerned, the islands were the Falklands and totally British. Indeed, to a Scotsman they were in many ways home from home. On an almost equivalent latitude, they bore a strong resemblance to the Outer Hebrides. They had the same combination of bog, hill and mountain, a predominance of deep browns and purples, and an indented coastline full of natural harbours and anchorages. The capital, Port Stanley, was like a new town of the Western Highlands with its rows of whitewashed cottages, and Government House resembled an Orkney manse. The main sheepstation of FIC at Darwin (so called because Charles Darwin, during a visit to Port Louis in 1834 with Captain Fitzroy in HMS *Beagle*, had ridden over to the locality and spent the night there) was simply a village of Scottish shepherds, some from the Western Islands, some from Ross and Sutherland, some from Inverness, and now joined by a sixteen-year-old from Dumfries.

Despite the company of compatriots and the similiarities to Scotland, William must have been fully aware that he was almost 8,000 miles away from home. Since being discovered, the Falkland Islands had been variously described by an eighteenth-century Jesuit as "this unhappy desert", by Samuel Johnson as "an island thrown aside from human use, stormy in winter, barren in summer; an island which not the southern savages have dignified with habitation", by the nineteenth-century writer on South America, William Hadfield, as "the best-suited spot in Her Majesty's dominions for a convict station", by Charles Darwin as "these miserable islands ... an undulating land, with a desolate and wretched aspect" and by no less than William Robinson, Governor of the islands from 1866-70, as "a remote settlement at the fag end of the world". The inhabitants were cut off from almost all humanity, and months would pass by without the welcome sound of the cannon from the government dockyard in Port Stanley announcing the arrival of provisions, mail, news and gossip.

The weather added to the sense of remoteness. On average it rained two days out of three, often in a drizzle. As Fitzroy wrote: "A region more exposed to storms both in summer and winter it would be difficult to imagine." Snow was recorded in every month, though it seldom settled long because of the exposure to the sea. The prevailing southwesterlies, cold and penetrating, reached Force 6 on at least half the days of the year, and an eighteenth-century report claimed that when the rare southeasterly blew, "Poultry are dropping, crippled, and those once lamed never recover but linger, and at length dye, with one side totally decay'd; the Hoggs are seized with the Staggers."

William had not expected an El Dorado. He had accepted a challenge: to learn all that there was to learn about sheepfarming and eventually to purchase a farm of his own. To all appearances he could not have chosen a more propitious time for his arrival, as a remarkable expansion was taking place in the sheepfarming industry on the islands. A few decades earlier, cattle rather than sheep had been the islands' mainstay. A handful of heifers and bulls landed by Frenchmen in the 1760s had proliferated into huge herds which, roaming wild on East Falkland, numbered about 40,000 head by the

late 1840s. Encouraged by the first governor, Lieutenant R.C. Moody, merchants imported gauchos from the mainland and a massive slaughter took place for the sake of the cattle's hides. Their extinction was only a question of time, the last ones being killed in the early 1880s.

Attempts were made to import domesticated cattle which soon proved to be unsuitable and uneconomic, but the pastures were ideal for sheep and the islands were fringed by abundant and nutritious tussock-grass, known as "the golden glory of these islands", which looked like small groves of low palm trees, one plant producing 200 to 300 shoots and some rising 5 to 7 feet high. First attempts to introduce stock in the 1840s, however, were disastrous as a result of rampant scab and a lack of labourers. Whole flocks were lost in the mountains, rams mated when they wished, lambs could not survive the colder months, and one flock was even destroyed by its owner's dogs. The breakthrough came in January 1852 with the foundation by Royal Charter of the Falkland Islands Company, which purchased land from Samuel Fisher Lafone, a merchant of Montevideo who in 1846 had been granted over 600,000 acres in the south of East Falkland, still known as Lafonia, and who had been one of the most active cattle-slaughterers. With its headquarters in Port Stanley and with interests in shipping, sealing and whaling, storekeeping, banking and merchant trading, FIC opened its first sheep station in 1859 at Darwin, some sixty miles west of Port Stanley at the end of Choiseul Sound, and imported fine-bred Cheviot and Southdown rams from Britain to be crossed with ewes from the River Plate. The result was a success. In 1847 there had been one hundred sheep on the islands and all of two bales of wool were exported. By 1862, the year William Halliday arrived, the flocks numbered 20,000, increasing to almost 65,000 by 1871 and to over 185,000 by 1875, between which years the islands exported over two million pounds weight of wool to Britain. By 1885 the islands would become financially independent of Whitehall.

These were, indeed, times of opportunity for a young shepherd. William lived for seven years at Darwin, soon acquiring the nickname "Skipper" from his tendency to take charge. To no one's surprise he was chosen to lead the shearing-gang. In 1869 he was promoted to Second Manager under a Mr Armstrong at Hillside House, halfway between Darwin and Fitzroy, a settlement south-west of the capital. Despite the strenuous work and periods of

6

7

8

9

prolonged isolation, life had its compensations. William lived in a simple but comfortable wooden house with its own poultry-run and kitchen garden. Outside the back door lay a limitless supply of peat, excellent fuel once the intricacies of the peat-spade and stove had been mastered. On his few days of recreation he could fish the rivers well stocked with trout, or hunt the many duck, geese and snipe. As long as the boggy terrain was passable he could ride over to other settlements such as those at Goose Green and Douglas, the latter named after its owner, or at San Carlos, named after one of the Argentine gauchos hired to kill wild cattle in the 1840s. He could share the mid-morning break, known locally as "smoko", or a rare "two-nighter" at which the first night was a whist-drive, the second night a dance. Each year he joined in the communal collection of berries, especially those of the "diddle-dee" shrub, welcome addition to a diet of mutton and vegetables eaten so perennially that it was known among the islanders simply as "365". In "Egging Week" he went in search of the delicious penguin eggs and those of the sheepfarmers' rival, the wild goose, six of which could eat as much grass as did a single sheep. Less humanely he would go gosling-chasing, trapping the weeks-old goslings among the long grass and tussock to make gosling pie. However atrocious the weather could be, there were some fine days, the sun shining on average more hours than it did in Scotland. He could visit the strange "stone-runs", rivers of rocks of random shapes that ran down some of the hills, the origin of which was a subject of conjecture. He could climb the high Mts Usborne (2312 ft.), Wickham (2056 ft.) and Smoko (1392 ft.) or in the sometimes remarkably clear atmosphere watch from the nearest clifftop the sight of thousands upon thousands of sea mammals and seabirds, the seals, sealions, gentoo penguins, rockhoppers and "mallemucks" (black-browed albatrosses). In the evening he could read a book or newspaper which sometimes reached the farms, or write letters that would take months to arrive in Scotland, or sit musing in front of the peat fire.

"Back home" in Dumfries, John and Margaret Halliday followed their son's career with some anxiety:

7 October 1870

William, my dear son,
    We received your kind and welcome letter and were happie to hear that you were in good health and pray for its continuance. We are all in good health at present. Grandfather and mother are both well and have been

expecting a letter from you for some time. They still have the cabs and horses and the town is getting better for the cab traid every year and housebuilding is going on very fast. There is a great difference in the appearance of the town since you left. You would scarcely take it for the same place for five years since. William, it is your mother's wish that you will be very careful. We are trying to do all in our power to save what we can for you and we hope you will be steady and saving, also be careful in your companions and choice of company as example goes a long way with everyone. So with this I enclose the portrate that I promised to send you, with our prayers for your saifty and well being and hopes to hear from you as often as convenient. Your friends are well at present. So, my dear boy, accept the kind wishes of your father and mother,

    J & M Halliday.

In 1870, William's parents could rest assured. Their son was progressing as a shepherd. He was depositing small but steady sums with FIC, albeit with no interest. True, there was almost nothing on which to spend his money, the few luxuries that arrived from the outside world being soon snatched up by the more worldly townsfolk of Port Stanley or by the garrison of thirty marines.

William did not regret his decision to leave Dumfries.

But during the next decade clouds began to loom on the Falkland horizon not only for the islands as a whole and for the sheepfarming industry but in particular for an ambitious shepherd in search of property of his own.

Due to their location, the Falklands had been important as a haven for ships en route to Cape Horn and the South Pacific. Since its foundation in 1845, Port Stanley earned valuable income from ship-repairing, especially after the discovery of gold in California and the establishment of a guano industry in Peru had led to an increase in vessels plying the South Atlantic. As soon as a ship was seen limping towards the islands, representatives of FIC and its only serious rivals, J.M. Dean & Sons, would row out into the harbour in a neck-and-neck race to get the business. With such a rugged coastline, the Falkland Islands had more than their share of shipwrecks, especially as there was only one lighthouse, built on Cape Pembroke at the entrance to Port Stanley in 1855. There were consequent disputes between the authorities and the islanders who were accused of ''conceiving the derelict of the ocean to be wholly the property of the discoverer'' and even of colluding with ship's-captains in the deliberate wrecking of their vessels.

By 1880, however, Port Stanley's importance was beginning to wane. Steam was fast replacing sail, and the newly founded settlement of Punta Arenas in the Straits of Magellan offered easier access and cheaper coal. The completion of the railway from the east to the west coast of the United States in 1869 greatly reduced maritime traffic in the South Atlantic, and by 1884 an officer of a British squadron stationed off the east coast of South America prophesied for the inhabitants of Port Stanley a future as bleak as that of "the surviving inn-keepers of coaching-days".

As a result of this decline, the islands became more dependent on sheepfarming, and thereby at the mercy of the fluctuations in the world price of wool. Despite the rapid growth in the size of flocks, there was an inevitable limit to the number of sheep which could be supported in an area approximately that of the Lowlands of Scotland. Settlements were established at all corners of both the main islands: at Hill Cove, Port Howard, Fox Bay, Port Stephens on West Falkland, at Douglas, Teal Inlet, Walker Creek, North Arm on East Falkland. The few smaller islands which could be used also had their settlements: Weddell, Saunders, Keppel, Pebble, Carcass. There was a serious risk of over-grazing. Because there were no fences, the sheep tended to flock to the best pastures and starved each other out. They also became strangely attached or "carrensed" to a particular area from which it was extremely difficult to remove them. The once-abundant tussock-grass was gradually used up. By the 1870s the disposal of surplus stock was already a problem, unsolved either by the establishment of a tallow-chandlery at Darwin in 1875 or by haphazard attempts at exporting mutton to the mainland. The farmers gave whole carcasses to their dogs or left them in the hen-runs, and, when killing en masse, they placed them in the small fields of barley and oats to attract the guano-depositing gulls. Sometimes they simply threw them over the cliffs.

William's own future also had its limits. In 1882 he was given the small pension promised under the original contract and was offered the chance to continue working with FIC. He agreed, but only on an annual basis, for there were serious drawbacks. In the first place he could not advance beyond the position of Second Manager as the more senior officials were appointed directly from Britain. More importantly it was impossible for him or any other shepherd to purchase or lease even the smallest piece of land. All available pastures on both the main islands had long ago been allocated to a few individuals and chiefly to FIC itself which owned ten large farms,

the largest of which extended from Darwin along Choiseul Sound to within a few miles of Port Stanley and by 1880 was stocked with over 100,000 sheep. Admittedly, even if William had been able to purchase some land, he would have later found himself in lengthy legal wrangling: no proper survey of the islands had been made, three different nautical charts were used as basis for boundaries, grants were assessed in nautical miles, though this fact was omitted from the grants themselves, and some were never officially registered.

With hindsight William could have been more careful about the terms of his original contract, but it is easy at the age of sixteen to ignore the details or to be misled. The stark truth was that although he had to a large extent fulfilled the first part of his ambition, to learn all that there was to learn about sheepfarming, he had no hope, at least on the Falkland Islands, of fulfilling his ambition to possess a sheepfarm of his own. In 1882 he was still only 37 years old. His savings and his pension potentially provided him with enough money to purchase some property. He could, of course, have returned to Scotland. But his father had died in 1876, the Gillespie business had been sold, he probably reckoned that his reasons for having left in the first place had changed little, and, above all, he would have considered such a return as defeat.

And it was now not only his own future that caused William concern. On 13 August 1874 he had married Mary McCall in what was one of the last "marriages by glove" recorded in the islands, in which the exchanging of gloves was considered sufficient proof of wedlock until an appropriate priest arrived to solemnize the union. A chapel belonging to the Wee-Free Church, a low branch of Presbyterianism, had been built at Darwin in 1860 but the Hallidays preferred to await the arrival of the official Anglican chaplain, Rev. Lowther E. Brandon, in 1877.

Like William, Mary was of pure Dumfriesshire stock. Her father, William McCall, was born in Durisdeer in 1824 and married Bridget Blanch Rae, a local lass, in 1845. Mary herself was born at Butterwhit in the parish of Dalton on 13 July 1854. Living at Hightae, near Lochmaben, the family were equally proud of their Border heritage: McCalls were known to have settled in the Dumfries area as early as 1500, had the strong clan motto "Ferio, Tego, Defendo" (I strike, I cover, I defend) and "had absolutely naught to do with the McColls". On his wife's death William McCall decided, at the age of nearly fifty to seek a new life for himself and his four daughters and, like his future son-in-law, signed a contract with FIC, arriving at Darwin in July 1873 with the young Mary, Jane, Agnes and Anne.

By 1884 Mary and William Halliday had seven children: Agnes Jane born on 31 July 1875, Mary on 20 July 1876, Willie on 24 October 1877, John on 20 April 1879, Margaret on 24 October 1881, Annie on 22 January 1883 and Archie on 2 October 1884. They were all authentic "kelpers", the name given to those born on the islands because of the enormous quantities of the large seaweed found on the shores. But what future was there for them on the islands of their birth?

William's misgivings were shared by his father-in-law and by several other shepherds including Jack Rudd, husband of Anne McCall, and William Douglas, George McGeorge, John Hamilton, John Scott and John Patterson. They were all determined to leave the Falkland Islands and the shackles of FIC.

But where to go?

The atlas was scanned. Discussions took place. And the shepherds' eyes and conversations fell time and time again on the most obvious place of all to which to emigrate: the mainland of South America, only 350 miles across the South Atlantic, where by all reports an enormous area of land lay waiting, virtually unsettled and with excellent pastures.

That area was called Patagonia.

2 *Patagonia*

WILLIAM BEGAN to make enquiries at Port Stanley from government officials, from sailors who had been through the Straits of Magellan or had touched on the coastline of the mainland, and from the islands' chaplain, Rev. Brandon, who had made some study of the history of the area. From all these sources it became clear that William was contemplating moving to a part of the world about which little of certainty was known.

Since Ferdinand Magellan's expeditionaries had become the first Europeans to land there on 31 March 1520, the word Patagonia had for centuries been used arbitrarily to describe the vast area between the River Plate and the Straits of Magellan. It was simply the last thousand miles of South America. By the nineteenth century its northern boundaries were accepted as the Río Negro and its tributary the Limay, its southern boundary the islands of Tierra del Fuego. Divided between Chile and Argentina, with the exact frontiers a subject of enduring international conflict, Patagonia consisted of three distinct geographical areas. The Pacific coast of Chile was precipitous, made up of a mass of islands and inlets with such heavy rainfall and dense forests that it was uninhabitable. To this day the only settlements between Puerto Montt and Isla Chiloé in the north and Isla Desolación at the western entrance to the Straits of Magellan are small colonists of foresters at the few accessible points. The eastern flank of the Andes was more amenable, with good rainfall and land rich enough to support cattle as well as sheep, but, due to the distances involved and the difficulties of transport, the colonization of the Andean region of southern Patagonia was likely to be only a corollary to that of the Straits of Magellan and the Atlantic coast as settlers eventually penetrated northwards and westwards. A feature of this area was a line of lakes along the eastern cordillera from Lago Nahuel Huapi in the north, through Lagos

GULF OF MEXICO
Longitude West from Ferro
ATLANTIC
OCEAN
TIERRA FIRMA
Carthagena
NEW K. OF GRANADA
POPAYAN
Quito
QUITO
Equator or Equinoctial Line
Para
PARA
MARAGNAN
Toupinambos
SIARA
RIO GRANDE
PARAIBA
FERNAMBUCCA
Tapuyas
BRASIL
SEREGIPPE
Truxillo
Santa Fe
Callao
Moxos
Chiquitos
RIO dos ILHOS
PORTO SEGURO
Caraba
Cartages
Xamucos
CHACO
Maragnu
Tropic of Capricorn
PARAGUAY
S. Catharine
SOUTHERN
Puelches
Buenos Ayres
Pampas
Rio de la Plata
BUENOS-AYRES
Araucos
Puelches
Chiloe I.
PATAGONIA
Falkland I.
Straits of Magellan
TIERRA del Fuego
Staten Land
C. Horn
PACIFIC OCEAN OR SOUTH SEA
OCEAN
Leagues 20 to a Degree
Miles 69 to a Degree
MAP
of
SOUTH
AMERICA.
For the Rev. Dr. Robertson's
History of America.
By Tho. Kitchin Senr.
Hydrographer to his Majesty.

12

Futalaufquén, Buenos Aires, San Martín and Viedma to Lago Argentino in the south. These lakes formed a natural dividing line between the Andes and the vast tableland or pampa that stretched to the Atlantic in a series of regular steppe-like plains. Huge rivers had once crossed from west to east carrying with them the torrents and débâcle of melting glaciers from the Andes. Many of these had completely dried up or become subterranean, the principal surviving rivers being the Rios Negro, Chubut, Chico, Santa Cruz, Coyle and Gallegos. The chief harbours were, from the north, Carmen de Patagones and Viedma at the mouth of the Río Negro, Puerto Deseado or Port Desire at the mouth of the Río Deseado, San Julián, with no river but one of the most protected bays on the coast, Puerto Santa Cruz on the Río Santa Cruz, Puerto Coyle on the Río Coyle and Puerto Gallegos on the Río Gallegos.

It is to this plain and these ports that the word Patagonia is now usually assumed to refer, and towards which William Halliday, as a sheepfarmer, was most attracted. But to which part? Much of the area was thought to be desert. When Charles Darwin ascended the Río Santa Cruz in 1834 he pronounced a famous condemnation on Patagonia: "The curse of sterility is on the land, and the water flowing over a bed of pebbles partakes of the same curse." True, the further east one moved from the rain-shadow of the Andes the more arid the terrain became. Cold currents chilled the warm air blowing in from the Atlantic and the prevailing southwesterly winds increased evaporation, reducing the quantity of what rainfall there was so that parts of central and eastern Patagonia could not even support the sturdy guanaco, local member of the camel family. But not all of the pampa was so infertile. South of the Río Santa Cruz, for example, ancient volcanic eruptions in the Andes had covered the area with sheets of basalt which spread to within fifty miles of Puerto Santa Cruz and almost down to the coast itself at the Río Gallegos. Here, the rainfall was heavier and, William was assured, the pastures richer. It was towards the south of Patagonia that he fixed his attention.

Just as the geography of the area was complex, so its history was mysterious. The usual explanation for its name was that Magellan's men found enormous footprints in the sands of San Julián and therefore christened the local inhabitants Patagones, from the Spanish for large feet. But there were doubts about this explanation. The footprints were so large, it was claimed, because the Indians

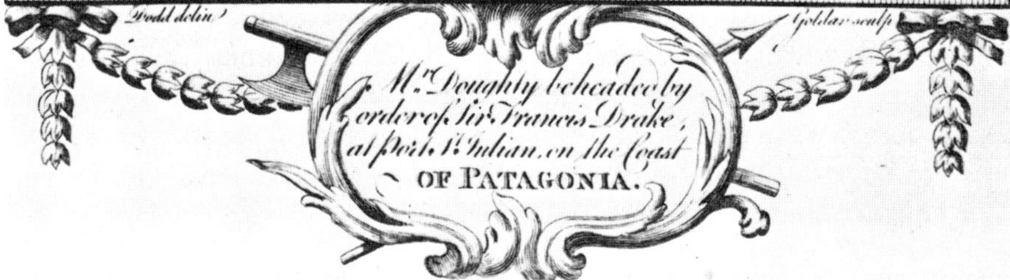

Mr. Doughty beheaded by order of Sr. Francis Drake, at Port St. Julian, on the Coast OF PATAGONIA.

wore boots made from guanaco-skin stuffed with straw, but there has been no subsequent evidence to substantiate this. Perhaps the Indians wore masks shaped like the heads of dogs and were therefore named after the Grand Patagón, a dog-headed monster which featured in a Spanish romance, a copy of which could have been among Magellan's possessions. Or perhaps the word originated from the Greek for "a roaring and gnashing of teeth", for the Indians certainly made plenty of noise. Or perhaps, in the light of agonies suffered by Magellan's men, by subsequent expeditions and by settlers both ancient and modern, the word should more appropriately be pronounced Patagonía, with the accent on the last syllable.

Whatever the reason for the choice of name, Magellan's expedition and other early navigators all returned to Europe with sinister reports. The names they gave to various geographical features show the impact made by Patagonia and its coastline: Deceit Island, Obstruction Sound, Useless Bay, False Point, Breaker Coast, Cape Shut Up. Sailors wrote of ferocious storms which lasted for days or even weeks. In Drake's expedition, for example: "The winds were such as if the bowels of the earth had set all at liberty; or as if all clouds under heaven had been called together to lay their force upon that one place." Hurricane squalls, nicknamed "williwaws" by eighteenth-century whalers, could compel a ship to anchor for weeks before continuing her journey, one anchorage being christened Forty Days. The prevailing southwesterlies drove a vessel repeatedly back into the Straits of Magellan as she tried to pull out into the South Seas and, as the Hallidays were to learn, would cause even a steamship to take eleven or twelve days in covering the 350 miles between the Falkland Islands and the Río Gallegos. Thick fogs resulted in the time-consuming dispersal of fleets and ships were further delayed by seaweed reputedly 1,000 feet long, and by whales. One fleet was "forced to looke well about us, and to wind and turne to shun the whales, lest we should sayle upon them".

The impressions made on early navigators were aggravated by the deterioration of their ships by the time they reached the Patagonian seas. Typical was the second expedition of Thomas Cavendish in 1591-2: "Our shroudes were all rotten, not having a running rope whereto we may trust, and being provided only of one shift of sails all worne, our top sails not able to abide any stress of weather, neither have we any pitch, tarre, nail, nor any store for supplying of those wantes." Ship's provisions had usually run low and the sailors' diet was likely to consist of "muskles, water and weeds of the sea, with a small reliefe of the ship's store in meale sometimes". With a lack of hygiene, dry clothes or fresh air between decks, the crews often suffered terribly: "Their sinews were stiff and their flesh dead, and many of them, which is lamentable to be reported, were so eaten with lice, as that in their flesh did lie great clusters of lice, as big as peasons, yea, and some as big as beanes." To find drinking-water was a major problem, as the Dutch navigator William Schouten discovered: "Our boat went to the north side of the river to seek for fresh water, but found none, for digging holes 14-feet deep, they found brackish water, both on high hills and in the valleys."

By the time they reached Patagonia, officers and crews were often ripe for rebellion. By a coincidence which helped to spread the infamy of the area, the first two men to circumnavigate the world both suffered serious mutinies in the port of San Julián, inappropriately named by Magellan after the patron saint of hospitality. Faced with discontent over the distribution of provisions and with growing scepticism about the existence of a passage through to the South Seas, Magellan was forced to execute the ring-leaders of a plot against him. Two of the mutineers chose to be marooned and were left behind on an island off San Julián with several sacks of biscuits and bottles of wine. Not surprisingly, they were never heard of again. More than fifty years later, in June 1578, the first thing Sir Francis Drake saw in Patagonia was the gibbet left by Magellan, and he was to use the same Island of True Justice on which to execute his close friend Thomas Doughty, found guilty both of inciting the crew and of conjuring the weather. With whatever chivalry Drake and Doughty acted, "each cheering up the other and taking their leave by drinking to each other as if some journey only had been in hand", and however masterfully Drake managed to control his dissident officers and crew, urging "the gentleman to haul and draw with the mariner, and the mariner with the gentleman", the message was clear: Patagonia did not welcome visitors. Drake placed a large grinding-stone over Doughty's tomb, carved with his name "that it might be better understood by all that should come after us". Two and a half centuries later, another British grave would join the site: "Lieut. Sholl, Officer of the Begle: died June 20 1828."

Early navigators also returned to Europe with a rumour from

which Patagonia would never be able fully to dissociate itself. It was said to be inhabited by a race of giants. When Magellan's crew came face to face with the Indians of San Julián they were astonished by their height, strength and voracity. One man was "fifteen spans" or over eleven feet high, another so tall that when he walked the Europeans could not keep up with him even by running, and when he danced, his feet sank eight inches into the ground at each leap. Lured on to Magellan's ship by a trick, two of the Patagons raged like bulls, ate basketfuls of biscuits and drank half a pailful of water at each gulp. Others swallowed rats without skinning them and pushed arrows down their throats without vomiting. As for the women, they were not quite as tall but very much fatter, with breasts "one half braza" or three feet long.

These reports were confirmed and even embellished by subsequent navigators. Drake's chaplain, Francis Fletcher, wrote: "They generally differ from the common sort of men, both in stature, bigness, and strength of body, as also in the hideousness of their voice." A Dutchman pulled the stones off an Indian grave and found bones that were 10 and 11 foot long. Cavendish measured an Indian's foot at eighteen inches. Others warned that the Patagons were "trecherous and of great stature; most give them the name of giants", and early maps and atlases contained illustrations of "gigantea" and legends reading "gigantum regio".

In the 1880s William Halliday had little difficulty in ridiculing these reports, but the existence of a race of giants was more credible to a sixteenth- or seventeenth-century mind. In the Bible the sons of Anak "came of the giants", the king of Bashan slept in a bed nine cubits by four, and Goliath stood at six cubits and a span, or over ten feet. Scholars from Pliny to St Augustine maintained that primeval man had been taller and stronger, and that the human race was steadily degenerating. Even an eighteenth-century French academician calculated that Adam had been 123 feet 9 inches tall, Eve only a little shorter, Noah a mere 27 feet, Abraham not more than 20, Moses 13, Alexander the Great 6 and Julius Caesar a puny 5 feet. Modern man would be microscopic had it not been for "the Christian dispensations" (Chambers). Mythical giants, offspring of the gods and humans, were commonplace in folklore: Giant Sigger kept his lair on the Orkneys, Herman and Saxie dwelt on opposite sides of Burrafiord in the Shetlands, and a small tribe of man-eating ogres had been destroyed on the hills of Dundee in the early

sixteenth century. London boasted its own giants, Gog and Magog. Wooden giants were normal appendages to royal pageants. Live giants were celebrities: John Middleton, "the Child of Hale", was said to be 9 feet 3 inches tall. Late seventeenth-century London could still believe that all Russians were inhumanly tall after Peter the Great strode down Regent Street. He was 6 feet 7 inches.

Yet there were, from the outset, doubts about the reports on the Patagonian giants. Navigators could, after all, say almost exactly what they wanted to say, knowing that it would be years or decades before their words were verified or disproved. The authors were usually not the leaders of the expeditions but members of the crew, who tended to exaggerate. Antonio Pigafetta, author of the most detailed of "Magellan's" reports, met Brazilian natives 140 years old, and it comes as a surprise that his captain presented one of the giants with an ordinary-sized sailor's outfit, unless, of course, it was for an infant giant. A member of Cavendish's expedition measured two Patagonian corpses twelve feet long and also claimed that the weather in the Straits of Magellan was so cold that one of the crew blew his nose off into the fire. Other reporters did not see giants for themselves but heard about them from rather shorter captives or, at best, saw them from a distance sitting in canoes or running up and down hills. Besides, the Indians tended to have disproportionately large heads and shoulders and wore loose guanaco mantles, making them look taller than they were.

By the second half of the seventeenth century, the reports were more realistic. One navigator found the Patagonians to be ordinary men, another put their maximum height at six feet, and a map of 1694 stated specifically: "Many people here of a common stature." By the time John Bulkeley and John Cummins reported back to England in 1741, the issue seemed to have been settled: there were no giants. But it was with their report in his hands that Commodore John Byron reached the coasts of Patagonia late in 1764 and set alight the controversy with claims as spectacular as any that had preceded them (and more in the vein of his grandson the poet, who indeed used parts of the narrative as source for storm-passages in his epic *Don Juan*). "Foulweather Jack", as he was known from his reputation for attracting abominable weather, sighted some Indians at Cabo de los Virgenes who were so tall that "the stoutest of our grenadiers would appear nothing to them". They were "the nearest to Giants I believe of any People in the World". The English press

Tuesday and Thursday.

THREE PATAGONIAN WONDERS

M. ETHAIR, M. BARTINI, & M. LAFFLAR.

From the New Strand Theatre,

Who will go through their Wonderful Feats of Strength,—shewing the ancient manner of the evolutions of the body, Pyramids, and Equilibriums, tending to display the first principles of mathematics. The surprising Feats of these Performers, which neither pen nor language can adequately describe, have filled every Spectator with wonder at their unsurpassed excellence.

On Tuesday and Thursday.

M. ETHAIR

Will go through his Wonderful

HERCULEAN FEATS AND GYMNASTIC EXERCISES

He will sustain himself in an horizontal position, and in that attitude will lift Two Quarter-of-a-hundred Weights with his Teeth, and One Half-hundred Weight in each Hand; also, bending backwards over a Chair, will lift a bar of Iron one hundred weight on the back thereof.

On Tuesday and Thursday,

A CHARACTERISTIC DANCE,

By M. ETHAIR, M. BARTINI,

announced that Byron had discovered "a new country in the East, the inhabitants of which are eight and a half feet high". The Royal Society earnestly discussed the news. Satirists set to work, notably Horace Walpole: "Captain Byron, being on that coast, saw a body of men at a distance, on very small horses; as they approached, he perceived that the horses were common-sized horses, but the riders enormously tall, though I do not hear that their legs trailed much on the ground." He asked if the Commodore had taken possession of Patagonia for the sake of the Crown so that His Majesty's full title could read: "George the Third, by the grace of God, King of Great Britain, France, Ireland and the Giants!" and wondered if some of the giantesses had been brought back to mend the British breed "which, all good patriots assert, has been dwindling for some hundreds of years". Byron and his crew remained strangely silent. A lieutenant at last confessed that "had the occurrence taken place anywhere else than in Patagonia, they should have set them down as good, sturdy savages, and thought no further about them". The next navigators to visit the area, Samuel Wallis and Philip Carteret, reported that, after measuring the tallest of the Indians, they had found only one who reached 6 feet 7 inches, the average height being from 5 feet 10 inches to 6 feet. By 1775 Thomas Falkner, a Jesuit priest who ran a mission in the hills of Tandil south of Buenos Aires, wrote: "The Patagonians are a large-bodied people; but I never heard of that gigantic race, which others have mentioned, though I have seen persons of all the different tribes of southern Indians." Darwin put their height at about six feet, though he conceded that they were the tallest race his expedition saw anywhere. George Chaworth Musters, the "Marco Polo" of Patagonia who spent a year in the 1860s wandering with a tribe of Southern Tehuelches, used measuring-rods and assessed the average height at 5 feet 10 inches. The Patagonian giants had at last faded into their proper realm of fantasy.

But legends die slowly. If the existence of giants had been disproved, Patagonia's traditional association with extraordinary size and with feats of exceptional strength or ferocity lived on. Shakespearean audiences had seen Caliban cower from the powerful Prospero who could quell even a Patagonian god:

> ... His art is of such power,
> It would control dam's god Setebos
> And make a vassal of him.

Impresarios drew fruitful comparisons: "Just arrived from Ireland and to be seen at the late Bird-shop, the corner of the Haymarket and Piccadilly, astonishing giants, whose height surpasses the Patagonian," and "Three Patagonian wonders: M. Ethair, M. Bartini and M. Laffler will go through their wonderful feats of strength." An eighteenth-century art critic complained:

> This year, of Picture, Mr. West
> Is quite a Patagonian maker —
> He knows that bulk is not a jest;
> So gives us painting by the acre.

And a group of Victorian entomologists were to be found examining an army of ants carrying off a "Patagonian centipede".

Patagonia was also said to contain giants other than humans. There were reports of birds with wingspans of over twenty feet and of hares the size of deer. In the cordillera there lurked strange creatures, some of them of a prehistoric nature. The Chilotes, from the island of Chiloé off the Pacific coast of Chile, firmly believed in the existence of the Trauco or Tranco which, possessing the form of a wild man covered with coarse shaggy hair, was said to descend from the forests and attack man and livestock. The nineteenth-century Argentine explorer Perito Moreno found a mummified human body painted in red, holding in its hand a large condor feather, also red. It was covered in short well-preserved hair and, more strangely still, part of its skull had been carefully scooped out as if to be used as a cup. No tribe of Patagonia was known to indulge in this sort of practice, and the skull resembled those of ancient tribes found in the graveyards north of the Río Negro, over eight hundred miles away. As interesting as how the body had arrived there was how it had managed to survive for so long in such a perfect state.

This led to the question of whether animals had survived from primeval times. Patagonia was probably one of the last places on the globe in which prehistoric animals had existed. In 1766 a skeleton was sent to Madrid from the region of Arrecifes, some 120 miles south of Buenos Aires. At first it was thought to be that of a human giant, but on closer inspection turned out to be that of mylodon, mastodon or megatherium. It was Darwin who in the 1830s realized the richness of fossil remains in much of the area of southern Patagonia: "It is a remarkable circumstance that so many different species should be found together; and it proves how numerous in kind the ancient inhabitants of this country must have been." From this abundance of fossil remains there emerged the inevitable and typically Patagonian rumour that some of these beasts still lived deep in the Andes. Those rumours would be put to the test.

In the late 1890s, surveyors who were carrying out preliminary research to determine the exact international boundaries between 51^0–52^0 South, found a piece of dried skin hanging on a tree: coarse, stiff and of a reddish-brown colour, it was encrusted with small bones and had evidently been exposed to the air for only a few months. The Indians thought the skin to be that of a cow or seal or even a Trauco, encrusted with pebbles. Perito Moreno thought differently. He managed to locate the cave at Caleto Consuelo in the inlet Ultima Esperanza on the Straits of Magellan. The cave was partly walled up in the front, and the interior was thick with dung and bones, suggesting that the animals had at one time been confined there by humans. Palaeontologists analysed the skin as that of the genus Glossotherium or Grypotherium, near relatives of the mylodon. The theory arose that these sloths had once been kept in a semi-domesticated state by the early inhabitants of Patagonia. The important question was how recently. The speculation was to reach such a point that in 1900 H. Hesketh Prichard, an English journalist, was sponsored to explore the Patagonian Andes. After a hazardous expedition, he found no evidence that any prehistoric monster had survived.

Despite repeated assurances that giants, both human and animal, did not exist in Patagonia, it was difficult for any prospective settler to dispel from his mind the niggling doubt that there might be some truth in the early reports. The Halliday children would later confess that they were at times very scared, and Grandpa McCall in his sterner mood had one infallible threat: "If you do some mischief, I'll call in the giants."

However evil Patagonia's reputation, the sixteenth-century European powers could not ignore an area of such strategic importance. By discovering a passage through to the South Seas, Magellan had found what was thought to be the key to the defence of the entire Pacific coast of America. If enemy ships could be prevented from using the Straits, they could also be prevented from attacking settlements to the north. In 1580 Pedro Sarmiento was instructed to study the feasibility of establishing a defensive colony and, reporting favourably, persuaded Philip II of Spain to finance a fleet of some twenty ships carrying a force of 3,000 men. By the time the expedition reached the Straits in 1584, the Patagonian seas had taken their toll, the fleet consisting of five ships and under 500 men. The survivors established two colonies, but plans to build sophisticated defences and even to stretch huge chains across the Straits were soon abandoned in the face of the more fundamental problem of survival. Supplies ran out. Many colonists died of hunger, thirst and exposure. Sarmiento set sail for Brazil and then for Spain in search of provisions and reinforcements but en route was taken captive by the English and, with no power in Spain interested in taking up the cause, within three years of its establishment the first settlement in Patagonia was extinct.

Only one man is known to have survived, thanks to the timely arrival of Thomas Cavendish, who in January 1587 met up with several of the Spaniards who had already abandoned the colony and were undertaking the Herculean task of reaching the River Plate by foot. The majority preferred to continue their futile journey northwards, but one Tomé Hernández was persuaded to come on board and he took the Englishmen to see "King Philipp's citie". Well-situated, it had four forts, four cannons, a church and a gibbet. But the place stank of death. The colonists had managed to survive for a while on a diet of mussels, limpets, roots, leaves and, from time to time, "a foule which they might kill with their peece". But then they had died "like dogs in their houses" or had fled inland to a probably more harrowing death. Cavendish named the place Port Famine: "The Spaniards which were there were only come to fortify the Straits, to the end that no other nation should have passed through into the South Sea, saving only their own; but as it appeared it was not God's will to have it."

And it would not be God's will for centuries to come.

The next Europeans to venture into Patagonia did not arrive for a military purpose but in search of wealth, because a rumour had broken out which was in its way even more sensational than that of the giants: Patagonia was thought to harbour a paradise. This fabulous city, Ciudad de los Césares, took its name from Francisco César, a captain under Sebastian Cabot and leader of one of several expeditions sent out from the early Spanish settlement of Fuerte Sancti Spiritú, at the confluence of the Rios Paraná and Carcaval over 1,000 miles north-east of Patagonia. César travelled into the unknown south-west and was said to have returned with heaps of gold and silver, gifts from an Indian chief. Finding his headquarters destroyed and abandoned, he retreated southwards and during the expedition some of his men disappeared under strange circumstances. It is easy now to guess their fate but the sixteenth century speculated in its own way.

The time coincided with several events: Pizarro's treacherous garrotting of the Inca Atahualpa in 1533 and the ensuing flight southwards of some of his followers carrying greath wealth, the disappearance of some of Sarmiento's colonists, and the steadily growing list of ships wrecked off the Patagonian coasts, many of whose crews suffered the same fate as that of John Chidley in 1589, recorded in Hakluyt: "And here at last sending off our boat to the island for the rest of our provisions, we lost her and 15 men in her by force of foule weather; but what became of them we could not tell." Out of a supposed union between the local Indians and some or all of the missing Europeans, there grew the legend of a fabulous city. Reports on its exact location and character varied as much as the reports on the size of the giants. It was opposite the island of Chiloé on the Pacific coast, or on Isla del Tigre in Lago Nahuel Huapi, or simply somewhere in the southern cordillera. It was surrounded by a high wall and a moat, and had one entrance with a swing-bridge. Its temples, ploughs, utensils, even furniture were all fashioned from gold and silver. Houses were studded with precious stones. The fields were full of cedars, oaks, orange-trees, poplars, and even palms. The city's inhabitants wore blue jackets, yellow waistcoats and three-pointed hats. They spoke an unintelligible language. The weather was perfect (in Patagonia!). There was no illness and, some claimed, no births or deaths. There was strict security and the inhabitants were bound to secrecy. They said that on the day that Ciudad de los Césares fell, the world itself would end.

The existence of such a paradise of riches and abundance was as plausible to sixteenth- and seventeenth-century Europe as that of a race of giants. There was the Gran Paytiti in Peru, the Mbae-Bera-Guazo in Paraguay, the Gran Quivira in New Mexico and, most famous of all, El Dorado itself, located at Lake Parima in Guiana or, more certainly, at Lake Guatavita, north-east of Bogotá in the Colombian Andes.

Fortune-seekers from Europe converged on South and Central America. In Patagonia they soon abandoned their glamorous dream in the face of a harsher reality, and by the 1620s the searches were at an end. But when Jesuit priests began trying to evangelize the Araucano Indians in the area of Nahuel Huapi in the 1660s, they could not ignore the supposed existence of a Garden of Eden. Fr. Nicolas Mascardi, the most influential, was given details of the city and sent messages ahead in seven languages, including Latin, Greek and Italian. Nevertheless, although his pastoral journeys probably took him as far south as the Straits, he failed to establish even the general whereabouts of the city.

William Halliday discarded the rumours of Ciudad de los Césares as readily as he did that of the giants. Yet, as late as 1871, Musters wrote that "the Chilotes and Chileans from the western side fondly cherish the existence of La Ciudad Encantada and the mythical people of Los Césares". And he told a story told to him by a Chilean who "knew a man who was acquainted with another who had heard from a third" of a party of woodcutters working opposite Chiloé, one of whom had lost his way in the mountains:

> Deep in the forest he had come upon a path, which he followed for some distance, till he heard the sound of a bell and saw several clearings, by which he knew himself to be near a town or settlement. He soon met some white men, who took him prisoner, and, after questioning him as to why he was there, blindfolded him and led him away to an exceedingly rich city, where he was detained prisoner for several days. At last he was brought back, still blindfolded, and, when the bandage was removed, found himself near the place of his capture, whence he made his way back to his comrades.

Gold certainly existed in Patagonia. The first gold found in modern times was in 1869 on the banks of the Río de las Minas, near Punta Arenas, and in the 1880s it was again found in sufficient quantities on Tierra del Fuego and at Cabo de los Virgenes for a gold-rush to take place, during which Julius Popper, a Romanian engineer, would join a long list of people who used Patagonia as a place to pursue vast personal ambitions, setting up a virtual dictatorship among the prospectors. In 1906 a gold nugget weighing 29 ounces was discovered at Río de Oro, Tierra del Fuego, and in the 1920s William Halliday's sons were to prove that primitive methods of exploitation could still produce an average of 4.5 pounds of gold a week.

After the disastrous fate of Sarmiento's "cities", Spain was reluctant to make any serious attempt at the colonization of Patagonia. Some tentative exploitation of the salt-lakes of San Julián in the seventeenth century merely resulted in a Paraguayan guard walking the long way to Buenos Aires to report that the installations had been destroyed by Indian attacks. The Spanish authorities were stirred into action, however, by the publication in 1775 of *A Description of Patagonia, and the adjoining parts of South America*, in which Thomas Falkner placed great emphasis on the potential of ports south of Buenos Aires. The book was also important as the first record of the culture of Patagonian Indians.

Born in Manchester in 1707 of Irish parents, Falkner studied medicine and worked in several hospitals in London until he fell ill in 1731 and was advised to take a sea-voyage. Embarking as surgeon on a ship carrying slaves from Guinea to Brazil and Argentina, he suffered a relapse in Buenos Aires but was so well treated by the Jesuit fathers of San Ignacio that, on his recovery, he decided to train for the Order and to remain in South America. At first he worked in Paraguay and north Argentina but in 1740 was ordered southwards to help Fr. Strobel, who had established a mission among the Indians of Cape San Antonio but was having trouble in competing with the local "wizards". For thirty years Falkner helped to run the mission of the Virgen del Pilar between the hill-ranges of the Vulcán and Tandil until, in 1768, all Jesuits were expelled from the Spanish Empires and Falkner returned to England. There he became chaplain to the political writer Robert Berkeley who, in conversations with the priest, realized the possibility of adding Patagonia, and perhaps all Argentina, to the growing list of British possessions. He persuaded Falkner to write a book, though it was Berkeley who gave it its political bias in a preface, and its title, Falkner himself not using the word Patagonia once in the text.

The implications behind the publication were serious enough for the Spanish to take notice and they decided to establish a series of defensive colonies down the Atlantic coast. These met with little success. A small fort built in Chubut at Bahía de San José in 1779 was abandoned after five years as a result of the usual Patagonian hardships, notably a lack of drinking water and an outbreak of scurvy. Another colony was established at Puerto Deseado in 1780, well equipped and consisting of an infantry officer, forty soldiers, seventy labourers or peons, a surgeon and a chaplain. But, within a year, its survivors were driven south by deplorable conditions and re-established themselves at Florida Blanca near San Julián, where they were joined by reinforcements who brought horses, mules, cattle, sheep, pigs and poultry. Receiving invaluable assistance from the local Tehuelche Indians, who provided them with drinking water and instructed them in hunting, the colony began to show signs of survival and even of prosperity. But in 1784 orders were received from the Viceroy of the River Plate that the settlement was to be abandoned, as the cost of upkeep was considered excessive and serious troubles in Buenos Aires required the presence there of all available militia.

The only colony to survive at this time was that of Carmen de Patagones in the very north of Patagonia at the mouth of the Río Negro. Dependent on cattle-raising and on the collection of salt, it was thrown into upheaval during the national rebellion against Spain but was re-established after Argentina's independence in 1816. The hostility of the Indians, however, prevented any exploration or colonization inland and it was not until the "pacification" of those Indians by ruthless campaigns, especially in the years 1879–83, that the government of Buenos Aires could contemplate large-scale colonization of the Río Negro. Further south a small colony had been established at the mouth of the Río Chubut in 1854, consisting of about fifty men and including an English sailor, Henry Libanus Jones, who had insisted to the authorities on the importance of defending the coasts. It survived for less than a year.

In 1860 the only official Argentine presence south of Carmen de Patagones was a solitary government agent on the Río Santa Cruz. Luis Piedrabuena, himself born in Carmen de Patagones in 1833, was granted a concession in 1859 to establish a trading-post at Isla Pavón, a small island ten miles upriver from Puerto Santa Cruz — a place so remote that he and his North American assistant William Clarke's first visitor from the outside world was the English traveller Musters, ten years later. Piedrabuena was responsible for representing Argentina throughout southern Patagonia, for preventing foreign sealers from trespassing on national waters and, above all, for wooing the Tehuelche Indians to the Argentine side in what was emerging as a serious confrontation over international boundaries with Chile. Charged with the distribution of rations and arms on behalf of the Buenos Aires government, he to a large extent succeeded in gaining the Tehuelches' confidence, but he realized that they were an apathetic people, less aggressive than the Pampas to the north and the Araucanos among the Andes, and more concerned with a regular supply of alcohol and tobacco than with territorial claims. Eventually they would be left to their slow extinction and would play little part in the division of land they once controlled.

Apart from Piedrabuena, the only other non-Indians in the Argentine area of southern Patagonia were a few missionaries. After the expulsion of the Jesuits there had been no attempted evangelization of the Indians for almost 100 years. But in 1853 a British warship, HMS *Vixen*, visited Bahía Gregorio and invited the influential chief or *cacique* Casimiro aboard. He expressed a wish to become a Christian and to convert his "ignorant" subjects, and this message was passed on to the Patagonian Missionary Society in London. Founded by Captain Allen Francis Gardiner in the 1840s, the Society's work was chiefly aimed at the Indians on Tierra del Fuego. First using Keppel Island in the Falklands as a base, a permanent mission was established by Gardiner near Ushuaia in 1855. After some initial success, he himself was killed by the Indians in November 1859, but the mission was continued by Rev. Thomas Bridges. As a result of Casimiro's interest, another small mission of two priests, Schmid and Hunziguer, was established at Puerto Santa Cruz in 1862. It was a failure. In order to secure regular visits by the Indians, the priests were forced to trade with them and this not only proved difficult because Piedrabuena's trading-post was already well established and better acquainted with the Indians' idiosyncrasies, but was frowned upon by officials of the Society. One of the missionaries went mad and their fate was sealed when a boat from the Falklands called in and exchanged rum and brandy for the Indians' fur blankets. A three-day orgy took place and the missionaries gave up all further attempts to instruct what they called "the little bright-eyed children".

During the nineteenth century, a few travellers managed to penetrate the interior of Patagonia, some of them British. While working from 1831-6 on the *Beagle* under Robert Fitzroy, successor to Philip Parker King and Pringle Stokes (the latter of whom had committed suicide in the Straits of Magellan when "prey to anxious fears"), Charles Darwin visited Puertos Deseado, San Julián and Santa Cruz during the project to complete a chain of meridian distances round the earth. In April 1834, he ascended the Río Santa Cruz to within sight of the Andes. Using three whaleboats and taking three weeks' provisions, Fitzroy, Darwin and twenty-five men found it impossible to row or sail against the strong currents, so the three boats were fastened together and the men, divided into two groups, hauled the boats upriver in $1^{1}/_{2}$-hour shifts. It took them seventeen days to cover 140 miles and, although only forty miles from the source, they decided to return, taking only one day downstream. On the expedition's return to England, Fitzroy wrote a two-volume narrative of the voyage and Darwin a third, in which he placed his infamous and far-reaching "curse of sterility" on Patagonia.

In 1848 a North American sailor, Captain Benjamin Bourne, was taken captive for three months by Indians near the Straits of Magellan and was forced to travel with them in the area of the Río Gallegos. An account of his captivity was published in London in 1853, in *The Giants of Patagonia*.

The next traveller of importance was George Chaworth Musters, who was born in Naples in 1841. Soon orphaned, he was brought up by his uncles, one of whom, Robert Hammond, had sailed with Fitzroy in the *Beagle*. After training for the navy at Gosport, he served in the Black Sea at the age of fifteen, and from 1861-6 as lieutenant on the sloop-of-war *Stromboli* off the coasts of South America, during which time he acquired something of a reputation after he and a midshipman, in a youthful prank in 1862, climbed the precipitous Sugar Loaf in Rio de Janeiro and planted a British ensign on the summit, much to the embarrassment of the Brazilians, who were unable to remove it for several months. He also bought land near Montevideo in Uruguay and started sheepfarming. In 1886 he was placed on half-pay and, with this and the profits from his farm, was able to realize his dream of travelling in the interior of South America, having first been inspired by Darwin's book. In 1869 he travelled to the Falkland Islands and from there to Punta Arenas. From April 1869 to May 1870, together with one Lieut. Gallegos, four soldiers and a mestizo guide, Aria, he roamed through Patagonia on a journey of over 1,400 miles, in the company of a group of Tehuelches, with whom he lived and like whom he dressed. On his return to England in 1871, he published *At Home with the Patagonians: A Year's wanderings over untrodden ground from the Straits of Magellan to the Río Negro*, an invaluable book which added to the scant geographical knowledge of the area and of the character and customs of the Tehuelches. It also contained the first reliable map of the interior of Patagonia. Musters became known as the Marco Polo or King of Patagonia, his journey described by the Royal Geographical Society as the most dangerous of any living man with the exception of Dr Livingstone. Considering that the Indians were reluctant to open their doors to the curiosity of white men in general, he achieved something of a diplomatic success, to such an extent that the Argentines suggested that he was in fact on a secret mission for the British Admiralty. After his return to England, he often preferred to sleep in the garden wrapped in a blanket in even the coldest weather. In 1873 he moved to Bolivia after marrying Herminia Williams of Sucre. In 1876 he returned to England and three years later, when preparing to take up the position of British consul in Mozambique, he died at the age of 38.

Another visitor to Patagonia at this time was the writer William Henry Hudson. Born in 1841 at Quilmes, near Buenos Aires, he contracted rheumatic fever at the age of 15 which prevented him from doing any manual work. Instead he roamed the pampa studying the many different birds of the River Plate. Employed by the Smithsonian Institute in Washington and the Zoological Society in London to collect specimens, he spent several months in 1871 in Patagonia, almost exclusively in the area of the Río Negro. His arrival was typical in that he was shipwrecked and had to swim and walk the last few hundred yards to the coast. In 1874 he left Argentina for good and went to England, where in 1893 he published *Idle Days in Patagonia*.

In 1877 a young Englishman, Julius Beerbohm, brother of the actor Herbert Beerbohm Tree and half-brother of the author Max Beerbohm, was inspired by Muster's book to travel from Punta Arenas to Puerto Santa Cruz. In 1879 he published the highly successful book *Wanderings in Patagonia or Life Among the Ostrich-hunters*. Essentially a bon vivant, Beerbohm was an unlikely person to take kindly to the rigours of travel in Patagonia. But on his return to England he so inspired another eccentric, Lady Florence Dixie, with his stories that in 1871 the two returned to Patagonia together.

Lady Florence, youngest of the six children of the 7th Marquis of Queensberry and therefore a member of one of the most influential families in Scotland, had already caused something of a stir in society. An inveterate tomboy, enthusiastic sportswoman and keen atheist, she had refused to grow her hair long enough to be "put up" when presented at Court, or to wear the obligatory lace and feathers; and, though darling of the "horsey" set and a favourite of the Prince of Wales, she was severely reprimanded by the latter when she turned up at Ascot in something resembling a nightdress. In 1875 she married Sir Alexander Beaumont Dixie, as solid and reliable as she was not. She soon grew restless and set out for Patagonia in 1879 with her husband, two of her brothers and with Beerbohm as guide, leaving behind her two-month-old son in Scotland, a decision much frowned upon. Her account of the journey, *Travels Across Patagonia*, published in London in 1890, was a bestseller at a time when women's travel books were in vogue. Perhaps tending to exaggerate, she wrote amusingly of how she nearly starved, was smothered almost to death by a pampas fire, strode knee-deep through rivers and slept in the open with a saddle as a pillow. After her return, she grew restless again and went to South Africa as war correspondent for the *Morning Post* during the Zulu War, after which she threw herself into the emancipation of women and the prevention of cruelty to animals, using her experiences in Patagonia to put forward her views in such children's books as *The Young Castaways or The Child Hunters of Patagonia* and *Aniwee, or the Warrior Queen*.

From the 1860s onwards, Argentina itself made some official exploration of the interior of Patagonia. In 1867 an expedition sent by Piedrabuena under the command of an English sailor, H.C. Gardiner, discovered the source of the Río Santa Cruz at Lago Argentino. In the 1870s Francisco "Perito" Moreno travelled from Buenos Aires to Lago Nahuel Huapi and from the Chubut Valley to Santa Cruz, discovering Lago San Martín with Carlos Moyano. Explorers such as Luis Jorge Fontana, Antonio Onetto, Ramón Lista and the Chilean Vicuña McKenna published accounts of their expeditions. But, on the whole, information on the area was still scanty in the 1880s.

Chile had been much more active in southern Patagonia. In 1843 a penal colony was established in the Straits of Magellan by President Manuel Bulnes. Fuerte Bulnes at first suffered the hardships of foul weather, lack of provisions and a plague of rats. It was moved to the more amenable Río del Carbon and renamed Punta Arenas or Sandy Point from the sandy promontory dominating the river. According to one English visitor, it was populated by a convict-guard only one degree less villainous than the convicts themselves. Two serious mutinies took place, one in 1851 and a second in 1877, after which the convict-station itself was disbanded. A small settlement remained, however, consisting of government officials, tradesmen, hunters, Indians who lingered on after their twice-yearly visits to collect provisions and barter wares, and, most importantly from the Hallidays' point of view, a small but steadily increasing number of settlers in search of land. In 1877 Dublé Almeida, the governor of the new Territory of Magallanes of which Punta Arenas was the capital, paid a diplomatic visit to the Falkland Islands in order to encourage settlers to the Straits. His hopes, and presumably those of his potential clients, were largely thwarted by the news of the mutiny, but Henry Reynard, an Englishman who accompanied Almeida as interpreter, took back 300 sheep from the Falklands to the Straits and placed them on Isla Isabel, leased to him by the Chilean government. This was the start of a rush in the settlement of the Straits. Some of the lessees were of Spanish origin: Cruz Daniel Ramírez on Islas Magdalena and Marta, Mario Marius and José Menéndez at San Gregorio, José Nogueira at Beckett's Harbour and Elias H. Bráun at Cabo Negro. Others came from Britain: Thomas Saunders, Arthur Waldron, Walter, George and Stanley Wood, and Thomas Fenton at Punta Delgada, Charles Fenton at Cabo Negro. They imported sheep from the Falkland Islands but, at this stage, few of the islanders themselves chose to move to the Straits. By 1884, the year that William Halliday and others first seriously considered leaving the islands, all the best "camps" along the Straits had been leased.

The government in Buenos Aires became increasingly alarmed at the Chilean domination not only of southern Patagonia, where the territory of Magallanes stretched unofficially as far as Puerto Santa Cruz and the town of Punta Arenas had a population of 500 by 1880, but also in the north-west, where cattle were driven back and forth through the Andean passes and where the Chilean militia, in its pacification of the Araucanos, was frequently trespassing on to territory claimed by Argentina. Faced with the difficulty of defending such a large and remote area, it decided that the best policy was to grant concessions of land to prospective settlers in the hope that any display of patronage would help the cause of national sovereignty. There was, in principle, no lack of bidders for those concessions, men lured by the very size of the area. In 1863 a German, Luis Bamberger, was offered a generous grant to establish a colony of 10,000 families. A group of Italians established a Patagonian Colonization Company and talked of importing 20,000 families. A Frenchman, Dr Brougnes, planned to convert the Indians into farmers. But these and other schemes died in embryo, victims of red tape and reality.

There was one notable exception. In 1865 a group of Welshmen arrived in Patagonia and founded the first settlement to survive in the area between Carmen de los Patagones and Punta Arenas. Their example was important to anyone else considering settling in Patagonia.

3 *Mhatagonia*

WILLIAM HALLIDAY had first heard of the Welsh colony of Chubut when he was still living at Darwin. Early in 1866, only a few months after the colonists had set foot in Patagonia, news reached the Falklands of the disastrous start to the grandiose dream of establishing a New Wales in South America. A sealing-vessel from Port Stanley had found the new arrivals in disarray at Golfo Nuevo and returned with two of the most disgruntled of them on board. Later that year the Governor received a petition from more of the colonists begging him, as nearest representative of the British government, to transfer the signatories to some other place. Urgent correspondence between the Governor and representatives of the British Admiralty in Montevideo resulted in HMS *Triton* being sent to investigate. The Welshmen were indeed in trouble. They had been reassured by their sponsors that the clothes they wore in Wales would be suitable for life in Patagonia. The *Triton*'s officers distributed sturdy shoes to the poorest, and her crew presented 1,000 yards of woollen cloth from which to make coats.

Lack of suitable clothing, however, was only one of a plethora of problems. The more William learned about the Welsh story, the more clearly emerged two warnings. First, he should settle as William Halliday, proud of his Scottish heritage but with no pretensions to establishing a foreign enclave on Argentine soil. Second, he should be prepared for the worst. The Welsh venture had initially been so riddled with mismanagement and lack of foresight that its survival and eventual prosperity were all the more remarkable.

The colony had been founded on a fantastic ambition. Ardent Welshmen like Lewis Jones, chairman of the original organizing committee, and Michael D. Jones of Bala, priest and teacher who had emigrated to Ohio for a time in the 1840s, grew increasingly alarmed

at the results of disorganized emigration from Wales, above all at the steady disappearance of the Welsh language, culture and traditions. What was needed was a place in the world to which Welshmen could escape from the destructive forces of the English language, the Anglican Church, the English landowner who was likely to evict tenants who did not vote for him at parliamentary elections, and a legal system which did not allow the Welsh language in courts of law even though most of the accused were monoglot Welshmen. Like many men before and since, the dreamers cast their eyes on Patagonia. Mhatagonia!

Contact was made with the Argentine government, who gave official blessing to the scheme. The site of the colony was tentatively chosen as the valley of the Río Chubut, largely because of Robert Fitzroy's "favourable" description after his visit in 1832, though, for the sponsors' benefit, his wild animals became "docile beasts peacefully grazing in lush pastures" and his driftwood was transformed into "standing timber". Public meetings were held throughout Wales. Promises were made: each settler would receive at least one hundred acres of land, five horses, ten cows, twenty sheep, two or three pecks of wheat, a plough and a number of fruit-trees. Great interest was shown. Two men were sent out to inspect the Valley for themselves and reported that, 50 miles long and 4 to 10 miles wide, it was not entirely dissimilar to the valleys of Wales. The colonists would be well out of reach of the English and, for that matter, out of reach of all humanity except for sporadic groups of Indians and a few prying Argentine officials. The Valley was christened Camwy. Final preparations were made. A handbook on Mhatagonia published in Liverpool confidently predicted in Welsh that "this single colony will grow to become a province of its own". But when the colonists at last set sail from Liverpool in May 1865, they numbered only 150, of whom 41 were women and 29 children. There were coalminers and slatequarriers, a tailor, cobbler, harnessmaker, a printer and a priest, but no carpenter, saddler, blacksmith, doctor, nurse or midwife. More importantly, for what was intended to be an agricultural colony, there were four farm-labourers but not a single farmer. Yet, as huge crowds wept and waved from the quay, those on board might have been forgiven for feeling a similar spirit to that of the Pilgrim Fathers. Everyone sang "God save the Queen" in Welsh. Then, as the Red Dragon was hauled to the masthead, the colonists chorused their own:

> "We have found a better land
> in the far south.
> It is Patagonia.
> We will live there in peace,
> without fear or treachery of war,
> And a Welshman on the throne.
> Praise be to God!"

But there were already ominous signs. Most of the colonists had spent their last money during a long delay in Liverpool and now could not pay for their passages at £12 a head. They reluctantly signed IOUs. The *Mimosa*, a small tea-clipper of 447 tons, was not the ship that had been originally promised. Makeshift accommodation had been erected on her deck, reached by a rickety ladder. She carried enough food and drinking-water to last six months as no one could confidently predict the length of the voyage. Each family was allowed to take six dogs and any number of chickens, ducks and geese. The scene was one of bedlam. Even now the journey proper had not begun. Just outside Liverpool the ship dropped anchor. Her papers were not in order.

After a three-month voyage full of sea-sickness and hymns accompanied on the harmonium, the colonists arrived off Golfo Nuevo on 28 July 1865 in their Sunday best. Lewis Jones and another man, who had been sent ahead to drive cattle and sheep down from the Río Negro and to prepare accommodation, had not yet arrived, so the first weeks were spent in huts hastily constructed from driftwood on the beach, or in caves among the cliffs where the first child was born, a girl called Mary, after whom the nearby hills were named. When Lewis Jones eventually arrived with the livestock, he found his compatriots hungry and annoyed: this was nothing like the place that had been described to them. The weather was bitterly cold, the landscape desolate and arid and the only vegetation were clumps of thorny bushes. He urged them to be patient till they saw Camwy, to which they would be moving in the spring month of October. But Patagonia was giving another characteristic welcome.

The Welsh had suffered their first casualty on the very afternoon of disembarkation. A young bachelor, Dafydd Williams, climbed the hills behind the beach to reconnoitre the area, and never returned. Years later his skeleton would be found near the "road" to Camwy. He had fallen victim to one of the risks of travelling even a

19

20

short distance into the interior of Patagonia: one could easily lose one's way, and then one's mind, and then one's life.

Within a few days, the sheep brought down from the north were lost on the pampa and the semi-wild cattle of the Río Negro were reluctant to be milked. Mrs Eleanor Davis, for one, more accustomed to the docile creatures of Abertefei, chased them, coaxed them and finally managed to get close to one that seemed quieter than the others, but it suddenly turned on her, and she, flinging her bucket at its head, ran off screaming, "The cows are bedevilled!" An exploratory attempt was made at planting wheat, but the soil was rock-hard and there was no source of irrigation except for puddles among the rocks formed by exceptional winter rains. The colonists decided to move to Camwy earlier than intended. The men travelled by land with the provisions, but their only guide was a copy of Fitzroy's chart and, with the bullock-wagons from the north proving cumbersome in the sandy soil, they were soon bogged down and lost. The women and children were to sail down the coast and into the Río Chubut in the *Mimosa* but, in Patagonian style, she ran aground and was lost, although without casualties. By a happy coincidence, another vessel sailed into the bay and was immediately chartered. After spending seventeen days trying to cross the bar, she at last deposited her passengers at the remains of the fort built by Libanus Jones ten years earlier, to which the men had eventually stumbled. There they established the colony's headquarters.

A feature of the fort was a moat that filled with water when the tide came in. The local Indians were reputedly scared of water, and the Welshmen were certainly scared of the Indians (even though no mention had been made in the original handbook of the still-ruthless Araucanos or of the attacks made on the town of Bahía Blanca as recently as 1859). The colony's minister, Rev. Matthews, later wrote: "If we travelled by night or slept outside on the camp, the shout of a bird was enough to cause great dismay. So we lived in a state of great fear, hour by hour, minute by minute, for several months." Ironically, when the Northern Tehuelches finally made contact they proved not only friendly but useful. They helped the Welshmen find freshwater springs, taught them how to hunt with Patagonia's most effective weapon, the *boleadora* — a type of three-balled sling — and later bartered meat and skins for the much-prized Welsh bread and above all, for bottles of rum and brandy which, much to the chagrin of the more devout churchmen, were to become an essential part of the Welsh trading-stores.

It was encouraging news for the Hallidays and other prospective settlers that the Welsh succeeded in living at peace with the Tehuelches. In the history of the colony there was to be only one serious incident: in 1885 four young Welshmen returning from an exploratory trip to the foothills of the Andes were murdered after being apparently mistaken for members of a rival tribe.

The official inauguration of the colony took place at the fort on 15 September 1865. A charter was drawn up for the running of the settlement and representatives were elected to act as a parliament. This compact document was one of the first examples of advanced democracy in the history of South America, even granting the vote to all women over eighteen years old, half a century before the enfranchisement of women in Britain. Each family had its own parcel of land and built a primitive house out of driftwood, using a box as a table and perhaps a cow's skull as a chair. A storehouse, built for holding communal provisions, was also used as a chapel and meeting-house. There was no cash as such but special sheets of paper were printed. The cabin from a wrecked ship was transformed into a school which the thirty children attended, mostly barefooted, collecting smooth stones from the hills to use as slates.

Seeds were sown, and prayers were said. But the crops failed, the first year from drought, the second from floods. Discontent and hunger grew in their place. Rowdy meetings were held. Some of the colonists simply packed up and left. At one stage in 1867, they all moved back to Golfo Nuevo ready to set sail north to the Río Negro, the furthest they could afford to go. Lewis Jones, who generally preferred to live in Buenos Aires, hurried back to Chubut in a last-ditch attempt to rally his protégés. The government in Buenos Aires, he assured them, had offered financial and agricultural support and, in any case, by the time they reached the Río Negro it would be too late in the year to sow seeds. They reluctantly trudged back to Camwy only to find that most of their homes had been burned to the ground by the Indians who, believing that the Welsh were leaving for good, had been indulging in their delight at the sight of fire. The colonists were at the end of their tether. After two years' suffering they were in a worse condition than when they had arrived. Sixteen of them had died, some by drowning, some from thirst and starvation when lost on the pampa. Forty-four had left. There had been twenty-one births. The only glimmers of hope were that, in a few weeks' time, fresh supplies would be arriving from Buenos Aires or that, in a few months' time, they would be leaving for the Río Negro.

It was during this wretched period that the colony was saved by what must have seemed a miracle but was, more accurately, the recognition of a basic principle of irrigation. By November 1867 Aaron Jenkins, formerly of Troedyrhiw, Merthyr Tydfil, was among those most adamant to leave. Having no plough and being too disconsolate to prepare his field with pick and shovel, he simply scattered his remaining seeds on the ground and lightly raked them over. One Sunday, reputedly while on his way to evening chapel, Jenkins noticed that the river, very full at the time, was at a higher level than his land. By cutting a small channel in the bank, he allowed water to run freely over the whole field, and then stopped up the gap. For the next few days the weather was hot and sunny, and the ground was soon covered with fresh green shoots. Several weeks later he repeated the process and by the end of February 1868 he had a fine field of wheat.

The colony was transformed. Soon every farm had its own ditches, and an extensive irrigation system was built. Not all the troubles were over, however. In 1869 the entire harvest was swept away by a freak storm and in some years the river fell too low to fill the irrigation channels. In 1899 there were to be such severe floods that deep-lying salt was brought to the surface and the quality of the wheat was permanently impaired. And, as far as Aaron Jenkins was concerned, Patagonia was to display its unhappy knack of wreaking private revenge on those who most actively defied it. In the early 1870s a fugitive from the penal colony in Punta Arenas turned up in the Valley. On such occasions the Welsh would hold the man in custody, bring him to trial and, if he could not justify his presence, send him to Buenos Aires on the first available ship. On this occasion Aaron Jenkins was chosen to accompany the fugitive to the judicial centre at the site of the old fort. On the journey the convict stabbed him to death at a remote spot and fled on his horse. The colonists were so outraged at the discovery of the death of their saviour that they chased the culprit for two days until they found him hiding among rushes by the river. They killed him there and then.

Finally, however, the Welshmen prospered. Vegetables and alfalfa were grown, butter and cheese were sent to Buenos Aires and the

21

22

wheat of Chubut began to win prizes at agricultural shows. Dams were constructed. A wharf was erected at Puerto Madryn, almost directly opposite the spot where the *Mimosa* had first dropped anchor, and a 40-inch gauge railway was built to cover the 47 miles from the sea to Camwy. New colonists arrived from Wales and Pennsylvania, and some of those contracted to build the railway decided to stay. By 1885 the whole of the main Chubut Valley had been developed into such villages as Gaiman, Bryn Crwn, Bethesda and Dolavon, and the population of the colony was over 1,600. In the same year several horsemen set out westwards on a three-and-a-half-month expedition to the foothills of the Andes, where they discovered a valley which they named Cwm Hyfryd, or the Delightful Valley, and established what is now the town of Trefelin.

In one important way, however, the Welshmen failed. Any hopes of establishing a New Wales in Mhatagonia were thwarted. Although the Argentine Minister of the Interior, Guillermo Rawson, had gone far beyond his government's guarantees in presenting substantial aid in the critical year of 1867 and although the colony was virtually self-governing for the first ten years of its existence, it was made clear from the outset that a foreign province on Argentine soil would not be tolerated. At the inauguration in 1865, the Welsh had wanted to name the place Caer Artur but Argentina had insisted on the name Rawson. The flags of the two nations were raised simultaneously. Indeed, the question of flags became a frequent subject of controversy: on one occasion Lewis Jones was arrested and taken to Buenos Aires when an irate Welsh wife tore down the Argentine flag and trampled it into the ground, and on another occasion the Welsh were seen saluting their own flag with such efficiency that a rumour spread in the capital that they were British soldiers in disguise. In their wish to escape the English, the Welsh had failed to realize the extent of Argentina's own nationalism. Every man born in the republic was subject to conscription and Sunday drilling, and the Welsh were absolved only from the latter, on religious grounds, after a visit by President Roca in 1899. Even the desire to foster the Welsh language was obstructed as the authorities insisted that children be taught in Spanish as well as Welsh. Any hopes of political autonomy were finally laid to rest in 1884, when Chubut was officially declared a territory of the Republic of Argentina.

The Welsh had proved to the world, however, that survival in the "deserts" of Patagonia was possible. In 1869 the editors of *The*

Standard, the English-language newspaper in Buenos Aires, had written: "From the very outset, the Welsh colony has undergone severe vicissitudes, as appears to be the fate of all new settlements." Twenty years later they were able to write: "The colonists have 3 flour mills, 8 threshing-machines, 70 reaping-machines, 6 pianos, 3 harps, a brass band and more than 100 violins.... Music is much cultivated, and Miss Lloyd-Jones is called the Patagonian nightingale."

William Halliday did not intend to move to Patagonia to grow wheat or to sing, but the Welsh success, especially in the peaceful co-existence with the Indians, must have given him encouragement.

4 A Futile Fish-Factory

IN 1870, five years after the arrival of the Welsh in Chubut, the Buenos Aires government granted a concession of land in Patagonia to a Frenchman, Ernest Rouquaud. The story of this small settlement was of particular interest to William because, situated at the mouth of the Río Santa Cruz, it was in the general area to which he himself was attracted.

Rouquaud had arrived in Buenos Aires in 1841 and established a large grease and oil factory in the suburb of Avellaneda. He made his fortune, but in 1870 yellow fever broke out in the city and, with many of the labourers dying, the factory stopped working. Walking melancholically along the banks of the Riachuelo, which ran past his defunct factory, he one day met Luis Piedrabuena, who was about to set sail for Patagonia. The Frenchman was told of unknown seas, unexplored lands and endless fish and seals. He made hasty negotiations with the authorities and was granted the right to open a factory on the Isla de Leones, some five miles upriver from the mouth of the Río Santa Cruz, where he would prepare dried fish, extract fish-oil and make soap. He set off with Piedrabuena, taking his wife and children, a bookkeeper, a mason, two carpenters, four personal servants and twenty peons, together with building materials, food for a year, horses, goats and hens, a cello, two pianos and eighty barrels of sugar for trading with the Indians. He established his home and factory on the island 100 feet above the river, reached by steps cut into the cliff.

By the time the factory was functioning late in 1872, Rouquaud found himself at the centre of an international storm. The Chilean government made it known through its embassy in Buenos Aires that it would not tolerate the establishment of Argentine colonies on or south of the Río Santa Cruz, claiming that, while boundaries were still under discussion, Argentina had no right to claim sovereignty over land in dispute. The governor of Punta Arenas, Oscar Viel, made matters as difficult as possible, reputedly using agents both to let loose the Frenchman's boats and to entice the peons to leave. Rouquaud appealed to Buenos Aires for help and President Sarmiento himself guaranteed naval protection, which arrived belatedly in the shape of one ageing gunship, the *Chubut*, in deplorable condition and with two cannons, both in the hold. Under orders to establish a *capitanía*, the Argentines built a few huts, erected a platform and flagpole, and hoisted the national colours. After three months, however, with no sign of Chilean opposition and with provisions running low, the *Chubut* returned to Carmen de Patagones and left the settlers to fend for themselves.

The odds were against the success of the factory: there were not enough boats, the strong winds made work precarious, the river dragged away the nets and there were fewer fish than had been anticipated. Trading with the Indians also proved unsuccessful because they not only abhorred eating fish but had little interest in sugar, their main desire being for alcohol. What alcohol arrived at the factory was at once consumed by the peons. At one stage, Rouquaud sailed to the Falklands and purchased a small herd of bullocks in the hope of starting a new venture, but the semi-wild creatures and a milking-cow presented by the Governor were soon lost on the pampa. Patagonia began to affect both the physical and mental health of the colonists, especially the bored womenfolk, who filled their time by helping to build a chapel, a school and even a theatre. Many of the peons packed up and left. When Rouquaud's elder son went on a spree to Punta Arenas and left his younger brother with strict instructions to look after the boats, the latter was drowned while trying to save several that had come adrift from their moorings.

But the colony persevered, and the vying nations continued to show interest. In 1873 a Chilean gunboat paid a visit, her captain claiming that he only wished to make use of the sandy beaches of the Río Santa Cruz on which to clean his ship's hull. But he was evidently under orders to impress the Frenchman, entertaining him lavishly, postponing his departure when Mme Rouquaud suffered a heart-attack and, at her subsequent funeral, urging his crew to carry lit candles and to give a solemn display of mourning. And then the Chileans left. Later in the year the *Chubut* limped back from the north but offered no opposition when a Chilean corvette called in to survey the river and to make an inventory of the colonists' possessions. The Chileans again left peaceably.

And so the naval games continued, each side waiting for the other to make the first aggressive move. Another Chilean corvette arrived, carrying on board Oscar Viel himself and several families of "colonists". The *Chubut* beat a hasty retreat upriver to Isla Pavón, not so much out of fear, perhaps, as out of shame at the lamentable state both of the ship and of her crew, who by this time had neither uniforms nor officers, the latter being either on sick-leave in Carmen de Patagones or on a hunting trip inland. The Chileans erected their own *capitanía*. The Indians joined in the fun, cacique Casimiro flying the Argentine flag, cacique Papón the Chilean. Rouquaud hoisted the French flag in a hopeless bid to display his neutrality. On Governor Viel's advice he sailed to Santiago, lodged an official complaint with the Chilean Foreign Ministry and with the Argentine Embassy, and was promised an indemnity by both, on condition that he abandon the colony. He reluctantly complied, returned to his factory and escaped with the remnants of his family, personnel and belongings, taking with him over seventy pieces of luggage, including one of the pianos and the untouched eighty barrels of sugar.

Meanwhile tensions mounted between Governor Viel and Piedrabuena. These erupted in September 1878 when a US ship, the *Devonshire*, began to collect guano from Isla de Leones with the full approval of the Argentine authorities. The Chileans immediately took her captive, an incident which caused demonstrations in both capitals and resulted in the Argentine Minister of War, General Julio A. Roca, ordering the military occupation of the Río Santa Cruz. Although Chile had the more powerful fleet, the authorities in Santiago decided to offer no further resistance on the grounds that, in the final analysis, the issue did not warrant the full-scale war that would have been inevitable. By the end of 1878 the Río Santa Cruz was under Argentine control.

As for Rouquaud, he vainly fought for compensation from both governments. The Chileans argued that, as his former settlement was now on Argentine territory, they had no obligation to pay. The Argentines showed little interest in his case now that the more important battle had been won. After seven years of haggling, the Frenchman persuaded each government to pay half of the indemnity originally offered. Despite their efforts after his death, however, the Rouquaud family never regained what they considered their legal rights to the property on which the factory had once stood.

Despite the events at the Río Santa Cruz, the border dispute was still by no means solved when William first considered moving to Patagonia in 1884. A treaty signed in 1881 proved unsatisfactory, the chief stumbling-block being how to interpret the exact physical features of the Andes, about which so little was yet known. A protocol was to be added in 1893 but it was not till 1901 that the matter was put to arbitration under a British Royal Commission led by Sir Thomas Holdich, who, after extensive exploration and patient diplomacy, persuaded the two sides to accept the principle of "divortium aquarum": in essence, that the sources of rivers that flowed into the Atlantic should be Argentine, those that flowed into the Pacific should be Chilean. In the 1880s William would be settling in a part of the world that was likely to be the scene of a serious confrontation and even of war.

Having gained a tenuous foothold in southern Patagonia, Argentina had an interest in establishing a more official and permanent presence. In January 1880, during the presidency of Nicolás Avellaneda, a resolution was passed in Buenos Aires to establish a colony at Puerto Santa Cruz. It was to consist of thirty families of at least three people, each of whom would be given a square league of land, 500 sheep, one milk cow, two oxen and materials for the construction of a wooden house with a zinc roof. The families would be obliged to pay back the price of fifty sheep after five years and, by that time, were expected to have constructed farm-buildings and to have cultivated a kitchen garden. When the colony was established later that year, it comprised only five families: Ibáñez, Garcia, Coronel, Clemente and Albarracín. After four years only Gregorio Albarracín remained, the rest having left as a result of suffering and official neglect. In July 1884 a similar settlement of nine families was established at Puerto Deseado. It was led by Antonio Onetto, an Italian engineer who had formerly been *comisario* to the Welsh colony of Chubut, but he died early in 1885, the sheep and provisions promised by the authorities never materialized and the colonists, with one gun between them, suffered misery and hunger. By September 1887 another official Argentine colony was at an end.

Meanwhile, as part of its endeavour to formalize Argentine occupation of Patagonia, the government of Buenos Aires in 1884 divided the area south of the Río Negro into four territories or *gobernaciones*: Río Negro, with its capital Viedma directly opposite Carmen de Patagones on the other side of the river; Chubut, with its capital at Rawson; Santa Cruz, with its capital at Puerto Santa Cruz; and Tierra del Fuego, with its capital at Ushuaia, southernmost town in the world.

The first governor of the Territory of Santa Cruz was Carlos Moyano. He had already been involved in some exploration of the area, ascending the Río Santa Cruz and travelling from Punta Arenas to Puerto Santa Cruz via the Río Gallegos along the coastal route. He established his headquarters at Los Misioneros, site of the defunct Anglican mission, and persuaded several of the former official colonists to join him, including Gregorio Ibánez, Saturnino and Cipriano Garcia, Manuel Coronel and Máximo Clemente. He had already come to the conclusion, however, that the best hope of populating his territory was to attract individual settlers rather than to make any further attempts at official colonization. To this end he pleaded with his government to draw up plans and conditions for the sale or lease of land and, like his Chilean counterpart before him, he saw the Falklands as the most obvious source of sheep and potential settlers. In 1885 he paid a diplomatic visit to Port Stanley and was impressed both by the state of the sheepfarming industry and by "the well-being, wealth and morality" of the inhabitants. He also returned with a wife, Ethel Turner, niece of the commander of troops on the islands, James Felton. As Moyano himself put it: "*Fui por lana, sin imaginarme que saldría trasquilado.*" (I went looking for wool, little thinking I would come home shorn.) On his visit he also detected some discontent among the shepherds. The land of Santa Cruz, he assured them, was similar to that of the Falkland Islands and, more importantly, the pastures were virgin.

William Halliday did not meet Moyano on this occasion. While the governor was in the Falkland Islands, William was already inspecting some of Patagonia for himself.

5 *Come, See for Yourself*

IN FEBRUARY 1885 William was given an unexpected opportunity to visit Patagonia as the result of meeting a German, Herman Eberhart, whose life was, in its unorthodoxy, typical of someone who had drifted to a part of the world from which it was difficult to drift much further. As a young man he had been sent by his father, a colonel in the Prussian army, to a military academy. Finding the place distasteful, he one day stated that he was going for a swim in a river, left his clothes on the bank and disappeared, presumed drowned. Five years later, however, he was brought to the notice of the German authorities in distant Peking, was deported, tried by a military court in which his own father was the president, and given twenty years' hard labour, the sentence being reduced to eighteen months on appeal against his father's bias. Once released, he emigrated to South America. He eventually settled in the Falklands and worked as a pilot, principally for the Kosmos Line of Hamburg which operated a service between Europe and Valparaiso and in 1880 signed a contract with FIC guaranteeing that one of its vessels would call in at Port Stanley at least once a month, thereby establishing the first regular and reliable mail-service the islands had known. On his journeys as pilot, ''Captain'' Eberhart witnessed the establishment of the first sheepfarms along the Straits and became increasingly impressed by the potential of the area. In 1884 he made enquiries at the *gobernación* in Punta Arenas and, on being advised that the best camps along the Straits had already been leased or reserved, he travelled on horseback northeastwards until he reached the pastures of the Río Gallegos. Pleased by what he saw, he tentatively staked his claim by pitching his tents at Chymen Aike on the Río Chico, a few miles south of the Río Gallegos itself. But he realized that, before leasing any land, he should be certain that wool-ships would be able to enter the estuary in safety as it would be

23

precarious, if not impossible, to load on to ships anchored off the main coast. Leaving his young companion, Willie Williams from Weddell Island, in charge of the tents and extra horses, he returned to Punta Arenas and from there to Port Stanley.

While in Montevideo earlier that year the British Embassy had approached him on behalf of a young Englishman, Hon. William Humble Ward, later Second Earl of Dudley, who was on a cruise round the world in his luxury yacht the *Marchesa* and was seeking someone to act as pilot and companion on a sporting trip round the Falklands and through the Straits. Meeting up with the *Marchesa* in Port Stanley, Eberhart accepted the invitation, on condition that in the course of the trip he should be allowed to try to enter the Río Gallegos estuary. Ward agreed, the *Marchesa* successfully crossed the bar and, having refurbished Willie Williams with supplies and himself with confidence, he continued his trip. He later told the Hallidays that, although he had declined to receive any fee from the young aristocrat, he was later handed an envelope which contained £100 "to help towards the farm". While holding discussions with FIC at Darwin, Eberhart met William Halliday and, learning that the Scotsman was among those keen to leave the islands, he praised the qualities of the pastures of the Río Gallegos, lush, virgin and there for the taking. He himself was intending to return to the area to make a closer inspection and to ride north to discuss with Governor Moyano the terms imposed on settlers. He invited William to join him: "Come, see for yourself." In February 1885 the two men set out from Port Stanley on the small steamship *Rippling Wave*, specially chartered for the voyage.

On arrival William at once preferred the look of the more undulating terrain to the north of the estuary. He purchased a horse from Eberhart, forded the Río Gallegos at the first crossing-point, Güer Aike, and rode the length of the north bank as far as Cape Fairweather out on the Atlantic coast. In this no man's land between pampa and cordillera, the hills were gentle and the pastures full of soft grasses, some of which he recognized from the Falkland Islands, others whose names he would have to learn: *cola de pinche, cola de zorro* and the invaluable *coerón*, which in winter protrudes through the snow and guides the sheep to the richer grasses below. There were no trees but plenty of shrubs, especially juniper and barberry. He came across at least two freshwater springs. Flocks of "ostriches", more accurately rhea, and herds of the strange-looking guanaco roamed the camp. He might have spotted the occasional fox or even the "lion" or puma which, he had already been warned, posed the most serious threat to sheepfarming in this area. A bird of prey, the carancho, sat menacingly, so it seemed, on the brow of every hill. And maybe, far above, a tiny spiralling speck revealed the presence of the condor, largest flying bird in the world. If this wildlife would appeal less to William in the future, for the moment it must have captivated a man who for the past twenty years had heard virtually no animal noises but the bleating of sheep and the cry of seabirds.

William climbed the highest hill, almost exactly halfway along the estuary. From there he reviewed the scene. There was a hint of the Falkland Islands and of Scotland in the landscape and the air. There were gullies or *cañadones* in which the sheep could shelter in bad weather. And wool-ships would be able to moor alongside the farm to load the bales that were already accumulating in his mind's eye. The hills themselves seemed suddenly alive with sheep.

But William could not afford to deceive himself. This was summer. In a few weeks or days or even hours, the picture could be transformed. And he and his family would be alone, except for Eberhart across the river, and Indians somewhere deep in the hills. He hazarded a guess that more settlers, including his own brother-in-law, would arrive in the area, but that would not be for months or years. Eberhart had told him of a rumour that the Argentine government were to build a sub-prefecture on the south bank of the estuary, in which case there would one day be no shortage of visitors. But William knew how long people and materials could take to reach the South Atlantic. Nor was he blind to the enormous difficulties of establishing a sheepfarm from scratch, nor ignorant of the infamy of Patagonia.

But his mind was made up. If the Hallidays of Annandale could march in their hundreds to the Crusades.... He would call the place Hill Station, a good, simple name, just close enough to the name Hillside to remind him of his days on East Falkland.

He returned to Chymen Aike and prepared to leave for the north to negotiate the purchase of his land. He later confessed to the children that this was the one occasion in his life on which, by way of celebration, he accepted the offer of "a wee dram".

It was when he climbed out of the Gallegos Valley on to the flat pampa that William began to see why Patagonia had left such a melancholy impression on virtually everyone who ventured there. He, Eberhart and "wee Willie" Williams were now riding on the fringes of a plain which would continue with increasing monotony and aridity for almost the next thousand miles. This was Darwin's "miserable country", Lady Florence Dixie's "grey, shadowy country which hardly seemed of this world", and Chilean explorer Vicuña MacKenna's "piece of petrified ocean, sterile, insensitive, solitary, silent and cursed", a part of the world which, however much he loved it, Musters conceded "wearied the eye with dreary sameness", and which Bourne called "bleak, barren, desolate, beyond description or conception — only to be appreciated by being seen".

The weather in summer could be surprisingly hot and sultry, with no protection against the sun. The feet soon began to swell in heavy riding-boots. Coats and rugs, so essential for the cold nights, were a cumbersome nuisance during the day. The limited drinking-water had to be carefully rationed, as a traveller could never be sure where he would find some more. It was easy to be fooled by what turned out to be the distant sight of salt-lakes. An unexpected problem was that of mosquitoes and flies whose presence, like that of parakeets and humming-birds, could come as a surprise to someone who had assumed all of Patagonia to be perennially windswept and cold. On the lake-lined fringes of the Andes the swarms of mosquitoes were so severe in the summer months that the huemul, a small but hardy deer about the size of a fallow, had to escape the lower grounds for the sanctuary of the snowline, and a human might have to resort to all kinds of stratagems: Lady Florence tied a handkerchief over her face, buried herself under her furs, lit some damp grass in the tent to smoke the enemy out, but finally resigned herself to "the passive endurance of the inevitable".

Some forty miles north of the Río Gallegos, just as the pampa seemed as if it would continue for an eternity, it was abruptly interrupted by the cliffs of a valley about four miles wide, through which flowed the small Río Coyle or Coig. The small expedition drank at the river, climbed out of the valley and were back on the relentless pampa. After three days William was probably more glad that he could have anticipated to descend into the valley of the Río Santa Cruz to the seat of Governor Moyano. With the governor himself still on his diplomatic visit to the Falklands, Juan Aubone, official accountant to the territory, had been left in charge and must have been happy to greet these rare visitors, especially as they seemed to be the very type of people that the Argentine authorities were trying to encourage to the area. Anglo-Saxon was thought to be synonymous with Progress. But the conditions imposed on settlers were still ill-defined. The Land Department in Buenos Aires was reluctant to commit itself to the direct sale of land because there was almost total ignorance of the topography of the area and the distance from the federal capital caused administrative delays. A temporary compromise had been reached, though this itself was only ratified later in the year: land in the Territory of Santa Cruz was to be leased for a maximum of ten years at a nominal rent of twenty pesos per square league, approximately four square miles. The tenant was at once to introduce at least a hundred sheep or goats, or twenty cattle or cows, per square league, and to build a house. Assurances were given that, when details for the sale of land had been agreed upon, the original tenants would be given first option to purchase the area they had stocked.

The Scotsman and the German accepted the terms. The rent was low and both men had already intended to stock their land with at least one hundred sheep per square league. They knew the pastures to be unsuitable for cattle, and goats had not featured in their plans! They had no reason to doubt that the Argentines would eventually honour their assurances.

William signed for twelve square leagues (over 30,000 acres) at Hill Station. Eberhart, a richer man, signed for two parcels of land each of eight square leagues, one at Chymen Aike, the other at Kilik Aike Norte, on the north bank of the estuary adjoining Hill Station to the west. Aubone confirmed that, as a result of an exploratory expedition authorized by President Roca, the construction of a sub-prefecture would begin at Río Gallegos in the near future. He guaranteed that, although inquisitive, irresponsible and often drunk, the Tehuelche Indians, some of whom were encamped near Los Misioneros, were not aggressive.

On the return journey, the weather changed for the worse and, in the face of a strong southwesterly gale, the going became slower. The men had few provisions and matters were not helped when "Willie dreamt he was hungry and ate more than his share of the rations". Fortunately Eberhart managed to shoot a young puma on the cliffs above the Río Coyle and, after five days, the riders descended into the Valley of the Río Gallegos.

Neither man wanted to delay. The *Rippling Wave* set sail for Port

Stanley on the first high tide. In the second week of May 1885, William rode into Hillside House, East Falkland, with the news his family had been waiting for. They were moving to Patagonia.

AFTER "THAT TRAGEDY" of 1 August 1885, William had two priorities: to make his family as comfortable and safe as possible, and to set off to the Straits to fetch the sheep without which the present catastrophe could not begin to end. At the thought of the task ahead he might have been reminded of a story told to him by his Grandfather Gillespie, and later passed on to his own children, of a Scottish laird who decided to plant some trees on his land: "On the first day of planting, he went to see how things were progressing and found the foresters fast asleep. 'What's the meaning of this?' he fumed. 'There is no hurry, sir,' came the reply, 'these trees will take half a century to grow.' 'In that case, there's not a moment to lose.' "

With their materials at the bottom of the estuary and with no timber in sight, the Hallidays strengthened their tent among the barberries, piling up shrubs all around and covering the sail with the skins of two old guanacos which had been picked clean by the caranchos. To start a fire they needed to find a type of shrub, *mata negra*, which would burn easily even when, as now, it had been covered in snow for weeks on end. There was no shortage of fuel, for the guanacos had the strange but convenient habit of depositing their droppings in communal heaps on successive days. The Hallidays found one such heap a few hundred yards up from the tents at the foot of Skunk Hill, so called because during the first days a skunk was seen scurrying over its brow. They lit the fire with Grandpa McCall's flint lighter, a treasured possession "at all times to be kept in an inside pocket". They picked leaves from the evergreen *té de pampa* or pampa tea, filled the large kettle with snow and brewed a beverage which, without the luxuries of milk and sugar, tasted insipid but, more importantly, was hot and nourishing.

The Hallidays were already learning that the prime rule for survival in Patagonia was to improvise, to wring every possible use out of the meagre assistance that Nature offered. They would discover, for example, that the common barberry or *calafate* bush could provide them with much more than protection against the weather. The Indians used shavings from its branches to mix with their tobacco, considered too precious a substance to be smoked on its own, and Grandpa McCall, who was almost always either puffing at a pipe or chewing "octaroon" tobacco, would later resort to this "blasphemous" practice when times were particularly hard. The Indians also extracted a yellow dye from the calafate roots with which to colour their blankets and mantles. In autumn the bushes would be thick with succulent berries for pies and jams. A French engineer on Holdich's expedition would even concoct *"un champagne à la calafate"*. Like many early settlers of Patagonia, the Hallidays would come to consider the calafate as a close friend. Andreas Madsen, a Dane who was to settle near Lago Fitzroy in the 1910s and became known as "Patagonia's Poet", wrote:

> Wherever I travel, midst cities' mad rush,
> Or at leisure o'er oceans roam,
> Dear memories go back to the Calafate Bush,
> My first Patagonian Home.

A tradition grew among the settlers that anyone who ate the berries of the calafate was bound to return to Patagonia.

With the prospect of abandoning his family and making the long return journey to the Straits, William must have regretted his decision not to transport sheep direct from the Falkland Islands. But that decision had been taken on good advice in Darwin, for on the few previous occasions when livestock had been brought to the mainland, it had often ended in disaster. The problem was that the only available craft were not suitable for carrying or landing animals, so the sheep were kept in the hold or in cramped pens on the deck. In rough weather, normal in the seas off Patagonia, the hatches had to be closed and many of the sheep were smothered to death. It was said that you could follow the tracks of schooners from the Falkland Islands to the Straits by the carcasses of sheep floating in the sea. There were times, it is true, when the animals were more fortunate. John Scott, a Scottish shepherd responsible for transporting the first sheep to Eberhart's farm at Chymen Aike in 1886, later wrote to the Hallidays of the precautions taken aboard the *Rippling Wave*:

> Just after embarking the sheep, we came to a nice green tussock island,

and dropped anchor there. The captain then assembled all hands with, I think, the exception of the cook, and told us to take knives and cut as many bundles of tussock as ever we were able, which proved the salvation of the sheep, for it was fifteen days before we reached Gallegos, where we landed the sheep in good condition. To save the fodder we tied up bundles of tussock round the pens and gave each a bottle of water every third day; so we were all kept busy, passengers and crew alike.

William gave final instructions to the family. Willie was to organize the construction of several corrals, for, as soon as the sheep arrived, they would have to be guarded all night against the splendid but deadly puma, "that cursed lion". Grandpa McCall would see that the ship's biscuit was scrupulously rationed and would kill, by whatever means possible, any of the abundant but elusive game. Mary herself needed no instructions. She had a babe-in-arms and six other children to feed and console. The family were on no account to wander too far from the tent or try to make contact with the Indians, for William himself wanted to be present when that first meeting took place. No, he assured his children, the lions would not attack, nor the Indians, nor the prehistoric monsters, nor the giants. There was to be a prize for the first member of the family to catch sight of William from the top of Square Hill, driving a large flock of sheep in front of him. He said he would be away a week or so, though he knew it could be much longer. Giving his usual unemotional farewells, he left his family in the care of Grandpa McCall and God.

He walked along the cliffs of the estuary with his two best sheep-dogs, Snap and Lassie, to pick up the horse he had left at Chymen Aike after the trip to Puerto Santa Cruz. Eberhart was still in the Falkland Islands making final arrangements to bring over his family and livestock, so wee Willie gave William his horse, after some argument as to exactly which one it was, and also presented him with the few provisions he could spare. It was to become an unwritten law among settlers that if anyone walked into another man's tents or house in search of food, water or shelter, he was unquestioningly given whatever was available. The rule applied to complete strangers and was sometimes taken for granted even if the owner was not at home. Two of Lady Florence's party raided the kitchen of an unknown farmhouse and helped themselves to what was to have been the farmer's breakfast, but when he himself appeared on the scene he was only a little surprised at the sight of "two dirty wild-looking men sitting uninvited in his kitchen". The hospitality was often inevitably meagre. Charles Akers, an English journalist who toured Patagonia as special correspondent of *The Standard* in the early 1890s, complained: "The food consisted of meat, either boiled or toasted in front of the fire or *asador*, some *fariña*, which more closely resembled sawdust than anything else, and, to follow the repast, some *maté*, or sometimes tea; if the latter, it was generally very weak, and without the luxury of milk." But a traveller learned to be grateful for the smallest of mercies. William would have been lucky to receive some dried meat, a few biscuits and some maté — also known as *yerba* or *verde* — a type of tea which was for most people in the south and, for that matter, on the whole Argentine pampa the only regular vegetable ingredient in their diet. (Traditionally the leaves were placed in a small gourd, hot water was added and the drink was sucked through a tube until dry. Maté was normally mixed with straw and other roughage because in pure form it served as an over-efficient laxative.) Julius Beerbohm wrote: "The country people drink it at all their meals, and whenever they have nothing particular to do, which is very often."

With no packhorse, William would not have been able to carry too many provisions even if they had been available. But at the thought of his long journey ahead without a rifle, and with the constant worry of his family back at Hill Station, he would have envied the lavish provisions carried by more fortunate travellers in the same area. Musters's two saddle-bags contained "a couple of shirts, and a jersey or two, a few silk handkerchiefs and soap, lucifer matches, writing-materials, fishing-lines and hooks, quinine and caustic, a small bottle of strichnine, a shooting-suit of tweed and a Scotch cap, and a most excellent pair of boots made by Thomas, a guanaco mantle, two ponchos and a waterproof sheet." His armoury consisted of "a rifle in case complete, two double-barrelled breech-loading pistols, hunting-knives, a small ammunition case of unfilled cartridges, and a supply of powder." Titus Coan, a North American evangelist who had lived among Indians on the Straits for a few months in 1833, travelled with "clothes, sheets and blankets, a small chest of medicines, a few books with stationery, two saddles with bridles and spurs, a few pounds of ship's biscuit and pork, a little bag of salt … axes, hatchets, files, fishing-tackle, saws, gimlets, augurs, hammers, handkerchiefs, garments, and several pieces of cotton and woollen goods, as presents for the savages." Lady Florence was furbished in appropriate style with "two small tents — tentes d'abri

— two hatches, a pail, an iron-pot for cooking, a frying-pan, saucepan, biscuits, coffee, tea, sugar, flour, oatmeal, preserved milk, a few tins of butter and two kegs of whisky." She had even contemplated bringing her whole retinue of domestic servants but eventually restricted herself to two, arguing that English servants "have an unpleasant knack of falling ill at inopportune moments". The party of Hesketh Prichard, the English journalist who was to go in vain search of a mylodon in the 1920s, carried "35 kilos of flour, 25 kilos of oatmeal, 15 kilos of sugar, 6 lbs of meat, 12 tins of cocoa; a spare change of underclothing, one tent, 50 rounds of 12-bore ball, 50 shot cartridges and 150 for a Mauser rifle — and 51 horses." William Halliday had one horse, two dogs, the clothes on his back, the few provisions from Chymen Aike and a handful of ship's biscuit. He had already experienced some of the hazards of travelling in Patagonia, but this second journey was different. He was alone, and it was winter. Winter! There was a saying that in some years in Patagonia there were only two seasons: a bad spring and a hard winter. August was a month of snow and a traveller could at any moment be overtaken and stranded by a snowstorm, leaving him short of provisions and causing innumerable personal discomforts. It became difficult to find fuel for the all-important night-time fire which not only kept him from freezing but also frightened off the pumas, which would reputedly never approach a flame. Musters once resorted to burning his tent-poles as he considered the fire more important than the tent. The bitter cold made it impossible to wash clothes as they became stiff as boards, and when fording a river, a traveller's boots could fill with water which froze.

In some respects, William was fortunate as the winter of 1885 was not, by Patagonian standards, a severe one. The journey might have ended very differently if he had been travelling in one of the infamous *inviernos malos* or bad winters of 1877, 1888 or 1902. In this last year, two young Englishmen with the Argentine-Chilean Boundary Commission, Edward Blinkhorn and Jack Lively, were riding from Comodoro Rivadavia to Lago San Martín via Lago Buenos Aires, a distance of over 600 miles, when they were suddenly overtaken by winter on the upper pampa. Lively wrote in his journal: "I left the first horse today. I wonder how many it will be before we see Lago San Martín. About fourteen or fifteen, I believe." His calculations were accurate. Unable to make any progress for days on end, the two men had to kill their horses one by one for food, and by the time they reached their destination over two months later, all that remained were the two horses on which they were riding, "and a handful of flour". William suffered no such catastrophe and, four days after leaving Hill Station, though tired and hungry, he rode safely into the farm or estancia of José Menéndez at San Gregorio on the Straits.

In buying sheep from Menéndez, on recommendations made in the Falkland Islands, William was dealing with a member of the family which would dominate the commercial life of southern Patagonia in the decades ahead. Born in Asturias, northern Spain, in 1846, Menéndez had arrived in Punta Arenas in 1874 and soon established the settlement's first important trading-house. In 1880 he leased an extensive tract of land along the Straits from the Chilean government, imported sheep from the Falkland Islands and stocked the first of his many estancias, on Bahía Gregorio. He opened links with Tierra del Fuego and eventually merged with Bráun y Blanchard (the company of his son-in-law Mauricio Bráun, son of a Jewish butcher from the Baltic), also based in Punta Arenas, to form La Sociedad Importadora y Exportadora de la Patagonia, which, by the turn of the century, would have a branch in every Patagonian settlement of any size and, by constructing bridges, schools and freezing-plants, by founding banks and by spreading electricity through the area, would play an important if often monopolistic part in the life of almost every settler.

For the time being, Menéndez was important to William Halliday as one of the only sheepfarmers in the area with stock to sell and with an interest in the export of wool. He bought 700 ewes and a dozen rams, the latter at £1 a head (pound sterling being the standard currency for southern Patagonia). He also arranged for one of Bráun y Blanchard's wool-ships to collect the first bales of wool from Hill Station "some time in January". With the transaction completed, provisions replenished and the thought of his family dominating his mind, William set out without delay on the return journey.

Despite years of experience with sheep, this was the first time that he had driven a flock through deep snow. It was vital to keep the sheep united, otherwise individuals became stranded in snowdrifts and had to be rescued or dug out, a time-consuming process. But Snap and Lassie were proficient at their job and, again, William was fortunate that this was not the year of a bad winter. In 1902 he would see the disastrous effects of snow on the flocks of Patagonia: further up the Río Gallegos one farmer would lose 60,000 head and the Hallidays would find some of their sheep so hungry that they were chewing the horses' tails.

Just as the winter of 1885 was not severe, so spring was approaching relatively early and William ran the risk of finding the Río Gallegos in full flood and impassable. In 1877 Beerbohm had been stranded for several weeks waiting to cross. Sleet and rain-showers were already turning the snow into slush and the streams into torrents. After living in the Falkland Islands, William reckoned he knew most of what there was to know about sleet and rain, that there was perhaps nothing more uncomfortable than, as Beerbohm put it, "the damp gradually creeping through your clothes, till it at last reaches the skin and chills you to the bone, while occasional rills of water run off your back hair and trickle icily down your shivering neck, till you are thoroughly drenched and numbed and cramped with cold". The soaking wet dogs had the unfortunate habit of trying to crawl under the blankets.

William also knew, however, that it did not pay to dwell too long on the details of personal discomfort. Each night he built as large a fire as possible. He learned the importance of keeping the provisions and leather objects well out of reach of those "sly marauders", the numerous foxes. There were two types of fox in Patagonia. The cordillera wolf or *canis magellanicus* was the larger, with a black and white back and reddish-brown flanks. Roaming the Andes, it was fearless of early travellers. Prichard found that it would approach to within a few yards during the day and even step over a sleeping man. The smaller greyish pampa fox or *canis azarae* was widespread throughout Patagonia east of the Andes. There were sporadic reports of it killing lambs but in general it caused little problem to the sheepfarmer, and even helped him by eating the eggs of the rhea and the geese, which competed for grazing. It was, however, a nuisance to the hunter because it could jump up as he was stealthily approaching his quarry and the dogs would give chase. Above all, both types were a constant pest for the traveller as they would creep up to eat almost anything. When a horse was tethered for the night, a fox could gnaw through the halter, set the horse free and leave the traveller stranded on the pampa. Their numbers would soon decrease, however, because of the popularity of their furs, although they were difficult to shoot as they were for ever twisting, turning and doubling back.

As the sheep approached the Río Gallegos, they showed their

uncanny knack of gathering pace when nearing even an unknown destination, so much so that the leaders had constantly to be driven back as the heavier and older animals lagged behind. During this time, William suffered his only serious mishap of the journey. He allowed his tired horse to graze for a while on Los Frailes Hills, south of the Río Chico. It took fright at the sudden appearance of a puma, broke its tether and escaped into the wilderness. William covered the last few miles to Chymen Aike on foot, to find that his horse had arrived there before him. It did not bear thinking about what might have happened had the incident taken place a few days earlier.

The Río Gallegos was already much fuller but not impassable. With the help of wee Willie, he cajoled his flocks across the river, knowing from past experience that the job was best done bareback as the horse would at one moment plunge into the water after the sheep, at the next rear up in natural reluctance, and a saddle could easily slip off.

Five days and nights after leaving San Gregorio, the flocks were grazing safely on the north bank, and a few hours later William was driving them along the cliffs to Hill Station.

He must have felt pleased, even a little proud, to have overcome the physical hazards of such a journey. But he would later confess to his children that, in some strange way, this and his former journey had as strong an effect on his mind as on his body. Though not a man given easily to meditating, he felt a sensation that was difficult to define or deny as he took a last look over his shoulder at the journey behind him.

As they grew older all the Hallidays would experience similar feelings. They could have pointed to several individual ingredients. Distance itself took on a new meaning. It was not only a matter of chilling statistics which a person would best ignore: that Patagonia was some fourteen times the size of Scotland, that the territory of Santa Cruz, the largest in Argentina, was over 180,000 square miles in area, that the distance from the Río Gallegos to Buenos Aires was some 1,640 miles, or that William had just made a round trip of over 200 miles merely to fetch some sheep. There was more to it. Because of the general monotony of the ride and the absence of all but Indian tracks, the precise number of miles covered became strangely irrelevant, even faintly ridiculous. Like all early settlers, the Hallidays began to use not the words mile or kilometre but the appropriately vague "league", which an authoritative dictionary defines diffidently as "an itinerant measure of distance, varying in different countries but usually estimated roughly at about 3 miles". Within Argentina itself, an official *legua* varied from province to province. Within Patagonia, it covered a multitude of distances. Musters complained that when he asked how far it was to the next settlement he was invariably given the reply, "A league," and Lady Florence discovered that a league could mean "ten miles one day and even more the next". The Hallidays considered it as "somewhere round about between four and eight miles", but they learned to treat distance with disdain.

By the standards of Dumfries and the Falkland Islands, William's journey to the south had been a long one. By the standards of Patagonia, it was trifling, a mere fifty leagues or so, and it paled in comparison with the journeys made by those who had come in search of gold, by the Jesuit missionaries, by Darwin, Musters, Beerbohm and Lady Florence Dixie, by explorers such as Moreno, Moyano and MacKenna, and by several of the Hallidays' contemporaries:

In 1888 five Scotsmen — John Hamilton, Henry Jamieson, John MacLean, George McGeorge and William Saunders — took almost two years to drive several thousand sheep and horses in a grand drive or *arreo* from Necochea to the Río Gallegos via Bahia Blanca, 1,500 miles.

During the bad winter of 1899, two German settlers at Markatch Aike on the Río Chico ran out of dip at a time when scab was rampant, built a sledge, trained dogs to pull it, travelled to the nearest trading-post, bought eight canisters of dip and covered the 34 leagues back to their farm in five days.

In 1904, an English settler, Jimmy Radbourne, rode on one horse through thick snow from the Río Gallegos to Punta del Este, a distance of over 50 leagues, in just over twenty-eight hours: "Halfway there he wanted to take a rest, but the foxes disturbed him so much and he found the bottle of brandy he had taken with him so bad that he continued without a rest — after pouring half the contents of the bottle down the horse's throat."

When commissioned to solve the border dispute between Chile and Argentina in 1901, Sir Thomas Holdich at first called for evidence but, being dissatisfied, decided to explore the terrain for himself. He covered over 800 square miles in just over two months.

In the 1890s a doctor, Victor Fenton, rode non-stop from the Río Gallegos to San Julián, a distance of 200 miles or 50 leagues, in an attempt to save the lives of a mother and her infant.

It was not just a question of distance and conditions. The traveller in Patagonia also had that niggling fear that he might at any moment lose his way, especially on the open pampa. On the "road" to San Gregorio, William had merely had to keep a strict southwesterly direction and the landscape, with its rocks and hills and increasingly thick vegetation, offered more landmarks than to the north. With no map or compass, however, his only guides were the sun, the stars and the tracks made by Indians which, except to an experienced eye, were indistinguishable from those made by guanacos. Unless he was careful, William would experience the chilling realization that he had been riding round in circles for the last few hours or perhaps days. It was even possible to lose the sea, as the English writer W.H. Hudson was to discover in 1893: "An hour's walk brought us to the hill. Climbing to the top, what was our dismay at beholding, not the open blue Atlantic we had so confidently expected to see, but an ocean of barren yellow sandhills, extending away before us to where earth and heaven mingled in azure mist." Any traveller would do well to follow the advice of the gaucho Martín Fierro, folk-hero of the Argentine epic written by José Hernández:

> See every day that your course you lay
> And watch that you hold it dead;
> Don't loiter or waiver or roam around,
> Just do your damnest to cover ground;
> When you sleep be sure that the way you go
> You're pointing with your head.

It was when a traveller lost his way that he was, paradoxically, most likely to discover himself. The macrocosm of Patagonia seemed to reflect the microcosm of the individual. Indeed, anyone, lost or not, soon learned the strengths and limitations both of his body and of his mind. For some people, the struggle was too much. Patagonia has a long history of suicides, such as those of Lieut. Stokes, officer in the expedition of Philip Parker King, and of an Argentine official sent to keep an eye on the Welsh colony in Chubut. Several settlers were to take the same drastic measure, either as the result of a specific catastrophe, such as a slump in the price of wool, a dispute with a neighbour, a bad winter, or from the cumulative effect of isolation, from the tedium of the ordinary day ordinarily passing against what was often a dismal backcloth. In 1900 Tom Jones, British consul in Punta Arenas, recorded:

> Whether it is the dreary and crude climate of Patagonia, or the lonely life in the camp after a day's work, or remorse after a bout of heavy drinking, I cannot say, but I have known well over twenty people who have committed suicide. Perhaps among the British it was a certain hopelessness at being in a hard country without any prospect of ever leaving it.

Patagonia often led people to walk the line between eccentricity and insanity. The Hallidays would meet many a "strange personality" in the years ahead, but many more would remain creatures of hearsay and gossip. The most famous of these was a Frenchman, Orllie-Antoine de Tounens, who had done nothing less than declare himself King of Patagonia.

Orllie-Antoine was most active in the north-west of Patagonia among the Araucanos Indians, but the story of his bizarre claims became well-known, and contained two warnings for any new settler: a foreigner should not meddle in national politics, especially in Indian affairs, and he should not allow Patagonia to lure him into megalomaniac visions.

Born in Périgueux in the Dordogne in 1825, Orllie-Antoine was a fanatical reader of the accounts of famous explorers and contrived a bold plan to reunite the seventeen Spanish American states into a Monarchical Confederation, of which he was to be king. Arriving in South America in 1858, he soon realized the impossibility of such a plan, but became another dreamer to turn his eyes on Patagonia. Between the Ríos Negro and Bío-Bío and the Straits of Magellan were an estimated 30,000 "noble Indians", divided into multifarious tribes whose only hope for the future was to be united under one leader. The Frenchman saw himself as that leader. He was an imposing man with long black hair, who wore a curved sabre and a French coat partly covered by a poncho, and his arrival in Patagonia in the 1850s coincided with a conviction among the Araucanos that the end of war and slavery would be brought about by the appearance of a White Man. The Moluches, a branch of the Araucanos famous for their hatred of the Chileans, received the Frenchman warmly and their cacique Quilapan was given details of the proclamations and decrees, of the coat-of-arms, flag and constitution of a country which would be called Araucania or Nueva Francia. Divided into departments and communes in the French style, it would have a Secretary of State, King's Council, Council of State, Supreme Court of Justice and Cabinet Ministers. All high officials would be obliged to take the oath: "I swear obedience to the constitution and loyalty to the King, and I promise to fulfil my duties with propriety and dignity." The State would pay for the sacraments, matrimony and funeral rites of all its subjects.

Quilapan approved, and Orllie-Antoine sent copies of his plans to the Chilean President, Manuel Montt, and to the press in Santiago, stating that he intended to invite all Patagonian tribes to join his kingdom. He visited Valparaiso to seek support from his friends and compatriots but, far from seeing the possibility of opening up new horizons for French colonization and culture, they treated the whole enterprise as a joke. The Chilean authorities, too, did not at first take him seriously; Orllie-Antoine merely accepted this as tacit condonement and returned to the Indians with enough money to pay for the hire of one horse and one peon, his constant companion Rosales. The Royal Treasury would soon put matters right! He visited the tents of another influential cacique, Levin, elected twelve of the tribe into a Parliament, gave one of his harangues on the Indians' rights of independence and promised that he would help all the Indians, on condition they made him King. Levin was impressed, and the twelve parliamentarians set out to spread the good news and gather more support. The promised White Man had arrived! The speeches were repeated and embellished, the new flag was handed to each tribe, a formal decree was written and Orllie-Antoine de Tounens was crowned King of Araucania. If his subjects greeted him or merely mentioned his name, they were to take off their headgear or, if not wearing any, raise their right hands to the skies.

But trouble was brewing. Mestizo interpreters were disgruntled at receiving no pay, traffickers in beads and alcohol were thrown into confusion and some of the caciques themselves became distrustful. The time also coincided with the concentration of the Republic's army with a view to a military occupation of "savage lands". When the King proclaimed that a contingent of armed Indians was to converge on the banks of the Río Bío-Bío to seek an advantageous peace with the Chileans, his interpreters denounced him to the authorities in Santiago. He was arrested and tried, first by the military and then by a civil court to which he objected on the grounds that one judge sitting behind a wooden table provided inappropriate justice for a king. He claimed that his aims were purely philanthropic: to build schools, to encourage agriculture, to stimulate the arts, and so on. When questioned about his original plan to reunite all the Spanish American states, he announced that Santiago would have been the capital of such a confederation. French diplomats in Santiago with some embarrassment suggested that the case be dropped immediately as the defendant was obviously not in full control of his senses. But after being given a medical examination and being declared sane, the self-claimed monarch was tried before a higher civil court and sentenced to ten years in prison. The Chilean authorities, however, remained uneasy. Orllie-Antoine's presence in the country was not only a source of friction between Chile and France but one of inspiration to the Indians. He received further publicity when he fell seriously ill in prison with dysentery and all his magnificent hair fell out. A compromise was reached: another civil court declared that, if not insane now, he had been insane at the time the offences were committed. He was to be placed in an asylum in Santiago from where any member of his family or French diplomat could release and repatriate him. The French consul of Valparaiso at once put him on a French warship bound for Brest.

That was not, however, the end of the King of Araucania. Once back in Paris, refreshed by the sea-journey and with his hair flowing again, he wrote his memoirs, received wide press coverage and published a manifesto promising jobs, abundant land and money to anyone accompanying him back to Patagonia. His appeal was to "the disinterested people of Old Europe, whose intelligence and hands are inactive because they have no place in the sun". He wrote letters to the Emperor and Empress, to the Foreign Ministry and to officials of every description, but did not receive a single reply. Embittered and penniless, he managed to return to South America in the middle of 1869, disembarked secretly in the bay of San Antonio and, with one private secretary, ascended the Río Negro to Choele-Choel. With the support of tribes who remembered him or had heard of his activities, he crossed the Andes and made contact with his greatest ally of old, the cacique Quilapan, who was still calling for a revolt against the Chilean authorities.

Hearing that the "madman" was back in the area, the militia organized a feast for friendly caciques and managed to extricate the names and whereabouts of Quilapan's supporters, offering a substantial reward to anyone who produced Orllie-Antoine's head. They then launched an attack against the rebel tribes, plundering their fields and burning their houses, but the Frenchman had heard of the price on his head, slipped out of Patagonia and made his way to Buenos Aires, from where, not receiving from the Argentine press the attention he considered suitable to his rank, he returned to

France. Despite the failure of his mission, he remained adamant and arrogant. He published his rights to the throne of Araucania, toured the provinces, exhibited his flag of green, blue and white stripes whenever possible and, by the end of 1872, was a familiar sight in Paris sitting majestically alone at the Café Mussard. He then planned to return yet again to his kingdom and, obtaining a loan from a London banker and using as an agent one J.M. Almeida, who preferred to be known as Viscount Palma, he sold bonds to finance his kingdom and purchased two ships in which to carry his retinue. The Chilean Ambassador in Paris objected to the Foreign Ministry, stating Araucania to be a province of Chile and denouncing the activities. Taken to court for fraud in the sale of bonds, excommunicated by the Pope as a freemason and refused a pension by the French government which he had claimed "for services rendered to France", the King returned to South America with a few faithful followers. Although shaved of his locks and beard and wearing huge black spectacles, while disembarking at Bahía Blanca he was recognized by a colonel in the Argentine army whom he had previously met at Choele-Choel. He was arrested, even though he claimed he was merely going to establish a colony and trading-house at Bahía Union near Carmen de Patagones. Persuaded by the French Ministry in Buenos Aires to vow that he would never again return to Patagonia, he was deported.

In Paris he continued to hold court with the few subjects and many onlookers in a dilapidated apartment in Rue Lafayette, with a stool for a throne. He offered decorations and gifts of nobility to anyone who gave him money. He sold specially minted coins. The Grand Chamberlain, one José de la Rosa, was in the habit of distributing a visiting-card which read: "The Duke of Rosenburg: member of the University of Esmirna and of various scientific institutions."

But Orllie-Antoine himself was now finished. Friends managed to find him the job of public lamplighter of Tourtoirac in his native Dordogne. On 19 September 1878, the King of Araucania, still a bachelor, died in the bed of a French provincial hospital.

In his will, he left strict instructions regarding a successor. The only relative found to be interested was a second cousin, Gustave Aguiles Laviarde. He did not have the same imposing personality, his wife "Queen Dona Maria" being perhaps the most attractive aspect of the new kingdom. He held a desultory coronation and never visited Araucania. However, he indulged in active propaganda by correspondence, carried decorations, held diplomatic banquets, gave a special envoy from Persia "The Royal Order of the Constellation of the South" and talked imposingly of how one day he would see his flag fluttering from the summit of the Andes blown by the winds of the pampas below. But his courtiers became depressed, journalists grew weary and the public had had enough. The second king of Araucania died of apoplexy in March 1902.

When the Hallidays heard told of the "mad Frenchman" they felt some admiration for him. For whatever reasons, he had at least treated the Araucanos with respect and had tried to uphold their identity in the face of aggression and exploitation. He gave them a voice which, however futile and short-lived, was more powerful than that of the Tehuelches of southern Patagonia who had no spokesman and would fade quietly and inexorably towards their extinction.

Patagonia affected some people with delusions less grandiose than those of Orllie-Antoine. When one of the priests of the Anglican Mission at Puerto Santa Cruz went in vain search of the Indians, he returned to find his colleague did not recognize him and had completely lost his senses. Others became misanthropic and cut themselves off from all humanity. George Greenwood, an ostrich-hunter, was known to have lived near Paso del Roble since the mid-1870s and was very occasionally to be seen in Punta Arenas. Lady Florence Dixie wrote: "He particularly affects that region and scrupulously avoids contact with his fellow-creatures ... he seems to live the life of a hermit. He has renounced the world and its vanities, even to the point of disdaining the ordinary rough comforts of the other inhabitants of the pampa. Clothed in the most primitive fashion, he roams along the slopes of the cordillera and, rather than make a trip to the colony to lay in a store of provisions, passes a whole year on a diet of ostrich and guanaco meat, pure and simple." In the depths of the bad winter of 1877, Greenwood was said to have made an "impossible" journey from Paso del Roble to Punta Arenas, during which he slept in the snow without a blanket and watched his horses perish one by one. Yet ten days later he was seen in Laguna Blanca. Jim Daniel, a naturalized North American, was another who lived completely alone with his six horses and twelve hunting-dogs, selling guanaco and ostrich-meat to passing ships.

Most famous of these loners was Ascencio Brunel, "the Wild Man of Santa Cruz", who was the subject of many a suggestion and accusation in the bars or *boliches* of Patagonia. He was known only to live "somewhere in the camp". One story had it that he and his brother both loved the same woman and that one summer, while the brother was horse-dealing in Punta Arenas, Ascencio rounded up the remainder of the stock and fled. But the woman in turn fled from him, and he lost his mind. Dressed in pumaskin from head to toe and armed only with the boleadora, he took to stealing horses but, unable to use or sell them, merely drove them into a sleepy hollow and killed them in their hundreds. In 1883 he was handed over to the police at Punta Arenas by Indians from the Río Mayo from whom he had stolen some mares, but he managed to escape and defied the law for almost fifteen years. In December 1896, hearing that he had been sighted in the area of the Río Gallegos, a search-party set off in pursuit and, on Christmas morning, heard the sound of a hymn among the hills of the Río Coyle. Shooting at random, they broke Brunel's leg and the breech of his gun, but it was several days before they caught him and took him to the police-station, by then built in the town of Río Gallegos. With the authorities unable to decide whether his offences should come under Argentine or Chilean jurisdiction, he was released, and, to this day, nobody is sure what happened to him. One rumour was that he died on the pampa soon after, having become sick from being introduced to a vegetable diet after so many years of living like a wolf. A German settler from far up the Río Gallegos later told the Hallidays another story:

> Brunel was still alive and well as late as 1897. In the spring it was agreed that my companion would go down to the coast with the cart for more material, including posts and wire to make a corral, and to get some provisions. Before leaving, we calculated when I could expect him back. When the time came, I got worried as he was two days overdue, and I thought his bullocks must have strayed on him, so I made up my mind to go down to meet him. Next morning I got up early and saw his four bullocks feeding along the banks of the river, which of course made me sure my thoughts about him losing his animals were right, and that he was looking for them in another direction. I only had to ride a day's march when I saw his cart. When I got up to it, to my horror I found him lying dead with a knife wound through his back, nearly all the provisions gone, also his horse and gear. That, of course, was the dirty work of Brunel!

Unusual behaviour was not restricted to these roamers of the pampa. There were to be plenty of personalities among the sheepfarmers themselves, though no one could be sure if they had come to

Patagonia because of their strangeness or if it was Patagonia that had made them strange. They were typified by "Mister Jack" Harris, an Irish woman who arrived with her husband in Río Gallegos in 1889. She was always dressed as a man, had short cropped hair, carried a large pipe and revolver and was, by all accounts, an excellent horsewoman, using a man's saddle, lasso and boleadora. One day she temperamentally packed her bags, left her husband in Río Gallegos and rode non-stop to Cerro Buenos Aires in the then-remote area of Lago Argentino. There she built a small house, established a farm, kept an immaculate garden and stocked the local rivers and lakes. But she did not take kindly to the routine of working with sheep and cattle, having the reputation of being the best-read person in Patagonia, and as suddenly packed her bags again, left for Canada on the first steamer out of Río Gallegos and was never heard of again. But stories of her lingered on, for example of an American who had worked with her at Cerro Buenos Aires for several months and, when on leave in Río Gallegos, told everyone how kind "Mister Harris" had been, even lending him some clothes. Mabel Halliday later recalled: "The American refused to his dying day to believe that Mister Jack was a woman."

To a non-Patagonian, the whole pattern of life could seem topsy-turvy. Akers found that, as all traces of civilization disappeared, everything was begun at the wrong end:

> If the people I encountered partook of a meal at which they had both soup and meat, they began with the meat and finished with the soup; if they happened to have any fish, they consumed this after the meat and before the soup. One day I noticed that the lid of a kettle was put on upside down and, asking the reason for this, I was told ... that such was the custom. Even when they lie down to sleep, they do not conform to the usual custom of mankind, but stretch themselves flat on their stomachs and cover up head and all with a blanket.

Suicide, insanity and eccentricity were the extreme results of the impact that Patagonia made on people, but each individual, even someone as sturdy as a Halliday, had to come to terms in his or her own way with what Darwin called "the free scope given to the imagination".

Some sought solace in alcohol:

> Alone, through spring and summer heat, till winter days draw in,
> Men count the steps about their shacks and drink consoling gin!
>
> (Madsen)

There were many excuses for picking up the bottle, not least the weather:

Blow here, blow there, it's little I care —
Come rain on my weathered pelt!
If I hit bad luck I tip my chin,
I take a good swig at my crock of gin;
If I'm wet outside I even up
With a sousing inside my belt.

As a teetotaller, William Halliday was to look with some dismay at the effect of alcohol on settler, peon and Indian alike. It was relatively easy and cheap to buy gin, rum, whisky, brandy or *caña* from a passing trader or at Punta Arenas, a free port where no custom dues were paid.

Other people found comfort in anything that broke the monotony, in objects which in normal circumstances would have been accepted as commonplace, but which, in the context of Patagonia, became lifelines to sanity. To a smoker, tobacco could seem a priceless luxury, especially if none had been available for some time. Grandpa McCall would have supported Beerbohm's eulogy: "Smokers will feel with me when I say my hand trembled whilst filling my pipe, and that, having lit it, I sat for a few minutes in a state of semi-ecstasy, enveloped in a fragrant cloud of the long-missed soul-soother." When stranded on the banks of the Río Gallegos, Beerbohm only managed to persuade his companion to attempt the dangerous crossing when the tobacco had run out.

For a gastronome, a single laurel-leaf could transform a stew into a memorable meal. Hudson raved about breakfast: "Cold boiled goose and coffee, often with no bread — it sounds strange, but never shall I forget those delicious early Patagonian breakfasts." Beerbohm ate an omelette aux fines herbes which "could have been prepared in a royal kitchen instead of in the Patagonian desert over a smoky greenwood fire, by the doubtful light of a few stars". And Lady Florence was in raptures when some officers from a German ship gave her an unexpected chance to supplement her diet of mutton with "asperges en jus which attained their delicate flavour under the mild fostering of a Dutch summer, pâtés elaborated far away among the blue Alsatian mountains, and substantial, though withal subtly flavoured, sausages from the Fatherland itself".

Some travellers gave names to their camps and devised extravagant menus in a bid to hold on to the lifestyle to which they had been accustomed and for which they often hankered. Prichard's party called one of its camps "Horsham Camp" and the menu at Lago Buenos Aires on Christmas Day 1900 read:

At 5 o'clock pm. Notice: Come early to get a good helping.
Menu:
Common or Garden Duff à la azuela
Condiment au lait Suisse
Grand Duff à la H. Jones avec muscatelles
Boeuf
Ostrich à la Patagonia — if you want it
Gigot de guanaco — order beforehand
Cocoa au lait ⎫
　　　　　　　⎬ Suisse
Thé au lait ⎭
Vieux cognac avec vulcanite
Plug tobacco
God save the Queen.

And a picnic was likely to be turned into a gala occasion:

Picnics in Patagonia, arranged by the Patagonian Picknicking Company on the most lavish scale:
On the free pampa!
Over glorious lakes!!
Through illimitable forests!!!
Ladies and gentlemen desiring to make this unique trip should communicate at once with the Secretary, Herr Bernardo Hahanson.
Unequalled scenery　　　Horse Exercising
Guanaco Shooting　　　Ostrich Hunting
. . . the engagement of picturesque heathen camp servants will be made a special study by the company.

Those who enjoyed reading were likely to snatch at any literature that fell into their hands, or to read time and time again any book in their possession:

A tallow dip, a book to thumb that's been thrice read before.

Musters's "library", for example, consisted of a novel, *Charles Dashwood*, which he was at one stage reading "for perhaps the twentieth time", and half of the "delightful" *Elsie Venner* by Oliver Wendell Holmes, which one of his companions had picked up to serve as wadding for cleaning his rifles and had reluctantly exchanged with Musters for a little gunpowder. Any newspaper would be scoured in search of startling items from abroad, or even for personal news that reminded the reader of a world that seemed of another planet. Jack Lively, who later became a friend of the Hallidays, once picked up a newspaper at Hill Station and remarked: "I see the old Regiment has been shaking up the Boers again," thereby giving himself a chance to tell for the umpteenth time how he had served in "D" Squadron, South African Light Horse, "Winston Churchill's regiment at Pretoria after his famous escape in the Boer War". If pen and paper were unexpectedly available, people wrote letters even if it was almost certain they would not reach their destination. Musters was assured by the cacique Casimiro that a missive he had written would be forwarded to the Araucanos who might then take it to the Río Negro: "Remote, however, as were all the contingencies, still it was a pleasure to write." Others waited eagerly for letters that would probably never arrive. At the sight of a ship sailing into the Straits, the missionary Coan leapt for joy at the prospect of hearing from his absent loved ones: "On the morrow, therefore, if the Lord will, one of us will hasten to the Strait." If no letter arrived, he indulged in effusive homesickness:

> No white-winged messenger bird comes over the wide waste of waters to tell us of our loved one's welfare. No morning greetings and evening benediction of "kith and kin" come to our ears. No voices arrest us but the harsh sound of the savages, the neighing of horses and baying of dogs, the roar of winds, and the rush and rattle of rain and hail.
> "The sound of the church-going bell,
> These valleys and hills never heard."

Others simply lay back, like Musters, "dreaming of home and taking a glass of sherry in dreamland".

Matters were worse when an individual fell ill. Although it was always unpleasant for a traveller or settler to be sick while far from home, in Patagonia the problems seemed magnified. The Hallidays would have agreed with Beerbohm:

> It is pleasant enough to roam over the pampa when you are strong and well, and can enjoy a good gallop after an ostrich in that pure, inspiring air, when the coarsest food seems delicious and you can sleep as soundly on the hardest couch as on the softest feather bed; but it is another thing when you are sick and in pain, and miss the darkened room, the tempting viands, the cooling drinks, and the thousand devices which make illness less trying.

It was to be a constant worry for the Hallidays that one of the children should fall dangerously ill or suffer an accident. In the early years, there was nothing they could do except hope that a ship called or that one of the family could ride to a nearby settler or to Punta Arenas in search of medical assistance. Even if an accident took place near a settlement, the journey there could be intolerable. Hudson was once shot through the knee by an amateur gunsmith playing with a revolver and, although only 36 miles from Viedma, the journey by bullock-cart "seemed to take an eternity".

Objects of nature in Patagonia took on a new significance. The Hallidays soon learned after a long spell at Hill Station what a pleasure it was merely to see a tree. Darwin felt "unmitigated joy" when at last he caught sight of the Andean peaks. Hudson rejoiced at reaching a river: "Never river seemed fairer to look upon, broader than the Thames at Westminster and extending away on either side till it melted and was lost in the blue horizon." The crowing of a cock excited in Beerbohm "a thrill of delightful sensation", and a herd of cows represented not only "milk and butter in perspective" but also "the softening influences of civilization".

To meet another human on the open pampa was likely to be a major event for both parties:

> ...a cloud of dust that means a passerby,
> Someone coming from somewhere, glad of an hour to lose.
> "Off saddle, man, have a maté, come and tell the news."

A traveller's horse instinctively gathered pace when drawing near to a settlement and the long journey was temporarily forgotten. Beerbohm wrote: "Fatigue disappeared as if by magic; the hills, hitherto so formidable, seemed to shrink into pygmy mounds; my pack became as light as a feather, and the rags round my feet seemed suddenly to possess the qualities of the famed seven-mile boots." Although the hoped-for town was likely to be a motley collection of huts, or "an undersized drinking-shop perched on the rim of the pampa" with tether-rail, a covered porch, an earthen floor, a long wooden counter scratched with graffiti and backed by a scaffolding of shelves bearing tins and bottles of indefinable commodities, with a few casks and cases on the floor, and one or two tables, stools and benches, a traveller could be sure that the occupants would be doing their very best to enjoy themselves, "concertinas and jackboots ringing in its galvanized iron-huts". There would be many a dance, singing-match and fight.

Yet there was probably a strange irony to this rowdy scene. Most of the revellers would soon be returning to the remote pampa, as ostrich-hunters, traders with the Indians, peons, shepherds or mere drifters. However enjoyable his time at the bar or *boliche*, each would confront again the isolation which was, to those who did not succumb to it, Patagonia's great and paradoxical attraction. Like many people before and after them, the Hallidays would find it both painful and invigorating, contemptible and irresistible. Darwin at one moment condemned Patagonia to the curse of sterility, at the next confessed "in passing over these scenes, without one bright object near, an ill-defined but strong sense of pleasure is vividly excited". He quoted some lines from Shelley. Hudson snatched at a newspaper with the instinct of a cat which, even when not hungry, pounces on a mouse, but he soon laid it down to listen to someone talking in the room, and eventually left the place without reading it at all. And why? "The answer was that I had drunk the cup of Nature, that my days had been spent in peace." A sense of loneliness had become one of solitude to which he returned time after time, "going to it in the morning as if to attend a festival, and leaving it only when hunger and thirst and a westering sun compelled me". Lady Florence Dixie was at first delighted to receive letters from home but soon found that they merely reminded her of "the responsibilities and pains of a world which but the day before had seemed so remote to us as if we had quitted the earth altogether". A letter from her game-keeper told "with interminable prosiness" how cleverly he had surprised the man whom he had so long and so wisely suspected of poaching, and "how C. had gone off to shoot big game in the Rocky Mountains" and " how D. had gone and shot himself". She spent much of her time in Patagonia grumbling: "To rough it may be all very well in theory, but it is not so easy in practice", and the Tehuelches were nothing but a nuisance, "wheeling their horses about the camp, careless of our crockery". In the next breath, however, she was comparing the place to a sonata by Beethoven, "severe, yet soft", claiming:

> Nowhere are you so completely alone, nowhere else is there an area of 100,000 square miles which you may gallop over and where, whilst enjoying a healthy, bracing climate, you are safe from the persecution of fevers, friends, savage tribes, obnoxious animals, telegrams, letters and every other nuisance you are elsewhere liable to be exposed to.

Beerbohm was delighted to meet another traveller on the pampa but the conversation soon began to flag, "being limited to occasional prophecies and conjectures as to when the weather might be expected to change for the better". Alone, he had discovered "a new sphere of free, fresh existence". He even found compensation for months without extravagant food, recommending that "every man who has pretensions to considering himself a gastronome should make it his supreme duty to give his palate a complete rest at least once a year. . . . He will return to his favourite dishes with that fresh zest and exquisite enjoyment which is vouchsafed to most people only in the balmy heyday of their schoolboy appetites". Prichard, while being disappointed by a Patagonian "town", conceded:

> It's a place which you hate and like at one and the same time. You long to get away from it while you are there, yet find yourself looking back sometimes and wishing to see again its vague streets and its degenerate agglomeration of humanity.

As William Halliday approached Hill Station with his newly-purchased flock, he was probably more delighted than he could have imagined at the sight of some of his children waving from the top of Square Hill. On the other hand, although used to twenty years' remoteness in the Falkland Islands, he had been strangely affected by his first journey alone in Patagonia. It was the start of an ambivalent relationship which he and his family would share in the years ahead.

8 *Pumas and Condors*

ON HIS return to Hill Station, William's first concern was, naturally, how his family had fared in his absence. They were safe and sound, but very hungry. Mary had breastfed Archie until her milk was badly affected by the poor diet, after which she could do little more than feed him with ship's biscuit mashed with boiling water, and hope for the best. The other children claimed his howling could have been heard on the far side of the estuary, and that it had helped to keep away the pumas and the giants. Then, just as the biscuit was running precariously low, the situation was saved. First, one of the guns which had been lost in "that tragedy" was washed up on the beach, and Grandpa McCall, cleaning it as best he could and using some dry powder and percussion caps from the box of tools and some shot fashioned from tiny pebbles, managed to shoot an ageing guanaco on the side of Square Hill. The meat was as tough as old boots, but no doubt delicious for the nearly starving Hallidays. Second, one morning a great quantity of fish was found stranded on the beach. Willie wrote in his diary for 16 August 1886: "The fish were up to our ankels." The probable explanation was that whales sometimes swam into the estuary itself in search of warmer waters and drove shoals of frantic fish in front of them, some of which swam to the very edge of the estuary and were beached by the falling tide. The Hallidays were to witness the phenomenon on only one other occasion when, in the 1920s, returning by boat from a trip to San Julián, "the propeller suddenly began to churn up hundreds upon hundreds of small dead fish at the entrance to the harbour". Mary, however, gave her own explanation: the event was "a miracle". Though never strict churchgoers or Bible-readers, the Hallidays believed in the works of the Lord — not that they were able to take full advantage of His help on this occasion. Without any salt, Mary was unable to cure the fish so she cooked as many as her family could

eat and packed some others in snow. The dogs, too, enjoyed the unexpected luxury, as on the preceding day they had been so hungry that they ate tallow candles washed up on the shore. There had been no sign of the Indians. There were plenty of puma-tracks in the snow but no sighting of the lion itself. Ostrich and guanaco had grown daily more inquisitive but remained elusive. Willie's corrals were at once put to use as William counted the flocks. Half a dozen ewes were missing, victims of snow-drift and puma. Agnes Jane wanted to know what prize she would receive for being the first to catch sight of her father from the top of Square Hill: the first motherless lamb born at Hill Station would be hers. That evening William killed one of the weaker ewes and the family ate a hearty meal to celebrate the birth of the sheepfarm. It would probably be a long while before they could afford to kill another sheep.

From that day on, and for every minute for the next few years, the Hallidays had to protect their flocks against "that cursèd lion". During daylight, the "shepherd of the day", who at this stage was either William, Willie or Margaret, would keep a scrupulous eye from horseback on the sheep as they grazed on the hills. Each dusk, which in winter was as early as 3 p.m., in summer as late as 11 p.m., the flocks were rounded up into the corrals, and the elder members of the family, with the help of the dogs and with gun at the ready, took six-hour shifts in an all-night watch.

The puma, or *felis concolor*, was a uniform sandy-brown colour except for the dirty white of muzzle, chin and inside leg. Four to five feet long, it could weigh up to 90 lbs and had a sinister growl and huge fangs. But the strength of its looks was in its eyes which, set in a relatively small head, were large, dark-brown and menacing. The Tehuelches called them "the eyes of the Devil". Named by Pizarro in the 1530s from the Inca word, the puma had once ranged the lengths of South and Central America, and also of North America where it was known as the cougar. Equally at home in mountains and forests or on savannahs and pampas, the "tygers" had been numerous around the sixteenth-century settlements of Buenos Aires, but as man encroached they were driven southwards. According to Darwin, their tracks were seen "almost everywhere" along the Río Santa Cruz in 1834. In the 1870s Hudson wrote that they still infested the settlements of the Río Negro and abounded in the hills of the Río Chico. One reason for their abundance in southern Patagonia was that the Tehuelches usually did not hunt them,

27

treating them as rivals maybe, but uninterested in their hide or flesh, both of which were provided by the rhea and guanaco.

In making enquiries about Patagonia, William had been warned of the number of pumas in the area and of the havoc they could wreak among a flock of sheep. What was of greater concern to him was whether the animal would ever attack a human. Never, Eberhart and others had assured him. Not even a babe in arms? No. A man on horseback was said to be perfectly safe. A man on foot was safe by day and at night had only to light a fire. At worst, the puma would show indifference. While searching for drinking-water in San Julián, two of Byron's crew came across a large "tyger" lying languidly on a *barranca* or steep bank:

> Having gazed at each other for some time, the men, who had no firearms, seeing the beast treat them with as much contemptuous neglect as the lion did the knight of La Mancha, began to throw stones at him; of this insult, however, he did not deign to take the least notice.

Just as with the reported friendliness of the Indians, the Hallidays preferred to find out for themselves about the puma and were never sure if it would carry out the threat its eyes implied. Despite all the assurances, instances of a puma attacking a man had been recorded. William Clarke, Piedrabuena's assistant on Isla Pavón, was almost killed when a puma grasped his neck from behind and tore the poncho off his back. The Argentine explorer Moreno was attacked on the banks of the Río Leona by one which "tore him under the chin with its claws" and one of Musters's companions, lost on the pampa for some days, vowed that he had been attacked several times in broad daylight and had killed pumas in self-defence. Perhaps the explanation was that the men had been wearing guanaco-skin ponchos and had been mistaken for the puma's favourite quarry. In the foothills of the Andes, where there was less available prey and where man was not so closely associated with rifle and boleadora, the pumas were known to be bolder and to reconnoitre explorers' tents even when a fire was burning. A Norwegian working with the Boundary Commission was seriously mauled near Lago Buenos Aires, partly because he was in an area never visited by man, partly because his fingers were so frost-bitten that he was unable to pull the trigger of his revolver. The poet Madsen claimed to have been attacked several times during his years in the Andes, but he confessed that on each occasion he was "the instigator".

William had already seen, from the incident in Los Frailes, how the sight of a puma could terrify a horse. Pandemonium broke loose when one jumped up in the middle of Lady Florence Dixie's cavalcade, "sending the mules and luggage horses stampeding away in all directions". A horse could seldom be induced to go to a place where a puma had been killed. And its fear was fully justified. However cowardly towards man, the "cursèd lion" was ruthless in its search for young horses, guanacos, rheas and that recent succulent addition to its diet, sheep. It had a frighteningly simple method of attack. Approaching with an immaculate combination of stealth and speed, it would leap on the prey's shoulders and draw back the neck with one paw until the spine cracked. Where sheep were concerned, a puma often killed for the sake of killing, slaughtering several but devouring only one, perhaps the result of the instinct derived from the days when it hunted in large packs or from sheer bewilderment at the sight of so much prey together on one occasion. If the sheep were in a pen, many would be suffocated in the panic that an attack engendered. The puma also had the unfortunate habit of selecting for its kill not, as might have been expected, the weak and stragglers but animals in peak condition.

The Hallidays had been warned that the puma was at its most dangerous during these months of September and October when, partly starved by the long winter, it searched desperately for food, hunting individually or sometimes in pairs. Not a single one, however, appeared for the first few nights at Hill Station. They had evaporated, as surely as game-birds at the start of the season in Scotland. But just as the Hallidays were beginning to relax their vigilance, a puma attacked the flock:

> Mother was about to change watch with Father and called him up to the "house" for a drink of tea before going down to the corrals herself at midnight. In the time it took them to finish the tea, a lion approached unnoticed. Hearing the hysterical barking of the dogs, Mother and Father rushed down the gully to see the cursèd animal slink away into the night.

They counted three dead sheep, all with their necks broken. Matters could have been worse. Madsen later recalled a puma that killed forty-five lambs and seven sheep in a single night at his estancia at Lago Fitzroy, Prichard one that killed seventeen sheep in two hours, and one settler, Percival Masters, had to leave Lago Roca in 1904

28

29

because of the number of sheep the pumas destroyed.

It was some small compensation that the Hallidays at least had some mutton on which to survive for a while. But this setback could not be allowed to happen again. Like all the sheepfarmers of southern Patagonia, they needed to exterminate the pumas from their farm as soon as possible.

Sometimes it was an easy matter. If chased on the open pampa, a puma first ran at great speed in a series of long leaps and bounds, but after several hundred yards it began to tire and was easily overtaken by a man on horseback who could despatch it with a rifleshot through the heart or skull. A hunter using a boleadora would lasso the puma once the weapon was entangled round its legs, and then drag the animal along the ground until it was unconscious. Sometimes it lay down as if dead as soon as it felt the noose around its neck, and then it was a simple task of killing it with a blow on the head. In forested areas, where the boleadora could not be used, the puma would be chased and forced to climb a tree by *leoneros*, long-legged terriers with a particular instinct for sport, specially bred and trained for the purpose. The puma would then be shot.

But the hunter did not always have things his own way. If there was any shrubbery in the vicinity, the puma would at once crawl into the very thick of it, safe from a direct rifleshot and from the boleadora. The hunter would have to pelt the bush with stones or even burn it down to make the puma run. It was when wounded and hiding deep in vegetation that the animal was at its most vicious and likely to kill any dog sent in to flush it out. In areas where there were caves, hunters sometimes carried a bottle of kerosene and a wick in their kits so that they could follow and discover where the puma was lurking. Mabel Halliday remembered once wriggling into a cave with a candle tied to the end of a stick. Others relied on courage and good fortune. After each snowfall, John Scott would look for puma-tracks near the caves above his estancia at Bella Vista up the Río Gallegos:

No tracks were visible in the recently-fallen snow, but I continued all the same. Scrambling among the rocks, I came upon some bushes. Suddenly a puma leapt out from under my feet, but was gone like a steak of lightning before I could take aim. At a distance of a hundred yards, I fired and discovered some drops of blood just outside a fissure in the rocks. Wanting to make sure my aim had been true, I climbed through, going down almost headfirst because of a slant in the rocks. From above, I suddenly spied a pair of malicious eyes gleaming at me. At that moment, I almost did not finish the job but slipped quietly out again. However, I decided to risk a shot; the big cat kicked and leapt in the air. I did not wait any longer. I have always wondered how marvellously quickly I escaped from those rocks. Having no further inclination for sport, I went home, and the next day returned to the scene of the kill. I found that the puma had fallen down a crevasse, and philosophically left him there to rot.

Like most sheepfarmers, the Hallidays resorted to the use of traps and poison. In the case of the latter, the pumas helped in their own demise. In former times they hunted in packs which could strip a carcass bare in minutes. When members of Byron's crew killed some guanacos and went back to the ship in search of help to carry them down, they returned to find that "the tygers had eat them, and had even cracked the largest bones to pieces". But now that the puma usually hunted alone, it would gorge itself on its favourite parts of the victim, normally the entrails and groin, and make a half-hearted attempt at concealing the remains under sticks and shrubbery ready for a future meal. If a farmer found these remains, or was guided to them by caranchos and condors, he could daub them with strychnine, taking care not to use too much of the poison lest the puma vomit it out of its system. He could then wait for the animal to return to its fate.

"Lion-hunting" therefore became an integral part of a sheepfarmer's life, and an enormous slaughter took place. Henry Jamieson, who settled at Moy Aike on the Río Coyle after the *arreo* from the Río Negro in 1890, made almost daily references to the puma in his diary:

> 5 *August 1896*: Tracked a lion for a mile; shot him at the lagoon; started three more; killed mine and gave C—— a hand to kill his; bent C's knife trying to spear him. Mc—— caught his alive and took him to his tent. Got home at 9.20. Damned cold.

> 16 *September 1896*: Mc—— poisoned another lion. He reckoned the lion killed him about 20 sheep. 23 lions killed this year.

In the 1890s two pioneers at Lago Argentino claimed to have killed seventy-three pumas in a week. It would take the Hallidays ten years before they finally "cleaned Hill Station of that cursèd lion".

By using poison, however, they inadvertently helped to remove from the Río Gallegos area another fine creature, for all too often the condors would discover a poisoned carcass before the puma had returned for its second meal. The birds were already quite scarce by the 1880s, being more numerous in the Andes, where they roosted and bred at heights of up to 16,000 feet, and on the Pacific coast, where in the summer months they scavenged the carcasses of sea mammals. At Hill Station weeks could pass by without one appearing. Then, quite suddenly, if a puma made a kill, or a guanaco or sheep lay dead or dying on the pampa, or a whale or seal were washed up on the beach, a few specks could be seen high in the sky. At the cry of "Giant-birds!" the Halliday children would scramble to the top of a hill to enjoy what had so thrilled Darwin: "It is truly wonderful and beautiful to see so great a bird, hour after hour, without any apparent exertion, wheeling and gliding over mountain and river."

Although not exclusive to Patagonia, the condor or *sarcorhamphus gryphus* was an apt creature for the land of the giants because of the sheer size of its wings and capabilities. In the right conditions it could fly distances of over thirty miles, only flapping its wings occasionally if there was insufficient lift in the currents of air. It could soar high among the cordilleras, rearing its young on a ledge of rock inaccessible to predators where the chicks or condorets remained for up to a year before making their first precarious flight. It was several years before the Hallidays were to see a nest, "a primitive, haphazard mixture of sticks and twigs", and they could never be sure if the condors seen at Hill Station had come from the Andes, from the precipices of Puerto Santa Cruz or, just perhaps, from the cliffs between Cape Fairweather and Güer Aike.

The sight of a condor in flight was all the more beautiful in comparison with the hideous appearance of the bird when seen at close quarters. Decidedly a member of the vulture family, the male had a flabby carbuncle or comb and beady pale-brown eyes, the female no comb but bright-red bloody eyes. Both had huge beaks, featherless wrinkled necks, almost uniform jet-black plumage and straight talons with a protruding middle toe. Only when the wings of a dead bird were opened was it possible to appreciate their enormous span.

Named by the Spanish conquistadores from the local Indian word *cuntur*, and known in English as contor, contur, condore, candore or condur, the condor had provoked reports almost as sensational as those on the human giants. Von Humboldt, the German naturalist and explorer, wrote of birds with a wingspan of nearly fifteen feet, and the English hymn-writer Samuel Wesley received as a birthday

present in 1827 "a prodigious bird ... above six feet in height ... whose wings, expanded, measured twenty-two feet four inches". The bird's existence, however, was not even acknowledged by European ornithologists until after a memoir given to the Paris Institute by von Humboldt in 1804. Some of the early reports were possibly exaggerated, for the largest bird the Hallidays ever saw was "a fully grown male weighing 20 lbs, about four feet long and with a wingspan of just over nine feet". The heaviest recorded was a specimen shot on San Gallan Island off the coast of Peru in 1919, weighing $26\frac{1}{2}$ lb. The average span of the modern adult male is 9 feet 3 inches.

The size of the condor helped it to acquire a reputation which almost led to its extinction: it was said to attack and carry off young livestock, especially lambs. Ulloa, an eighteenth-century captain of the Spanish navy, reported from Peru:

> Observing, on a hill adjoining to that where I was, a flock in great confusion, I saw one of the condors flying upwards from it with a lamb betwixt its claws, and then at some height drop it; then, following, took it up and let it fall a second time, when it winged its way out of sight, for the fear of the Indians who, at the cry of the boys and the barking dogs, were running towards the place.

Admiral Anson, eighteenth-century circumnavigator of the globe, claimed that condors would "kill a calf and devour a great part of it". It needed only a little stretching of the imagination for the birds to be credited with the harassment of travellers in the Andes and with the kidnapping of human babies.

The condors were certainly voracious. Though equally capable of surviving without any food at all for over a month, one bird could eat a calf and a sheep in a week. The powerful beak was able to tear the toughest horsehide or the skin of a sea-lion. Prichard once left a dead huemul unguarded for a few minutes while he went to fetch his horses, and took the precaution of tying a red handkerchief to the carcass to scare away predators, but when he returned the condors had already picked the huemul clean. They were certainly known to hasten the death of an animal already lying moribund on the pampa. At Hill Station they were sometimes seen converging on an old guanaco that had fallen on its back and was vainly trying to stand up. They were likely first to peck out its eyes and the flesh from under its cheekbones, and then to tear the rest of it to pieces.

This reputation made some Patagonian sheepfarmers believe that the "giant-birds" posed a threat to flocks, especially at lambing time, and on many farms near the Andes, sheepdogs were trained to bark whenever the birds lurked overhead. A great slaughter took place. Some tribes of Indians, too, killed the bird, which represented the triumph of the Indian spirit over the conquistadores and, according to Inca legend, each day carried the sun on high from its resting-place on sacred Lake Waynaquocha. It was also a symbol of good health: its heart was eaten raw as a cure for epilepsy and heart disease, its bones as a remedy for rheumatism, its stomach for problems of the breast, its roasted eyes to sharpen the sight, its blood to lengthen life, and, if its feathers were placed under the blanket, a man would be immune from nightmares.

Condors were sometimes easily killed. Once gorged, they became heavy, slow-witted and somnolent. Sometimes they slept in trees before returning to their rocky ledges and a hunter could place a noose around their neck. An effective way of catching them was to put a carcass inside a small enclosure of tall sticks on a level piece of ground and wait until the birds were well fed and heavy, when they needed more space in which to take off, and so were trapped inside the enclosure. Another method was for a hunter to hide under the bait in a covered pit, ready to grab the legs of the condor through a first-size opening, at which moment other hunters would run up to destroy it, though the birds' wings were sometimes powerful enough to knock down anyone who approached.

There is little concrete evidence for the farmers' belief that the condor killed and carried off healthy young livestock. Although the birds' talons were suitable for tugging and tearing, they were unable to grasp and therefore could not carry anything in flight. Their reputation was almost certainly a false one. At Hill Station, certainly, a condor was never seen to carry off a lamb. The Hallidays came to respect the bird, not only because it helped clean the camp of unwanted carcasses but mostly for the wonder of its flight. With regret they realized, too late, that in poisoning the puma they were exterminating the condor.

They felt less concern, however, at the fact that they were also poisoning the ubiquitous carancho or chimango. Although common through much of Argentina, it was typically Patagonian in its strangeness. Resembling a bird of prey, particularly a hawk, it nonetheless had the habits of a vulture and fed on carrion. Darwin wrote: "A person will discover the necrophagous habits of the

31

caranchos by walking out on one of the desolate plains and lying down to sleep. When he awakens he will see on each surrounding hillock one of these birds patiently watching him with an evil eye." It would also wait for ewes to give birth in order to eat the afterbirth and in case any lamb was stillborn or weak; it had the peculiarity of walking sedately around its dead or dying meal, sweeping the ground with its wings. It even picked at the sores on a horse's back. It was certainly able to grasp carrion in flight and one reason that the condor had acquired the same reputation may be that it was sometimes mistaken for a large carancho. Its numbers had increased throughout Argentina since the introduction of large herds of cattle to the pampas, and the Hallidays were to see a similar increase with the expansion of the flocks. It was never popular at Hill Station and was usually referred to as "that scavenger".

BESIDES GUARDING THEIR flocks against pumas, the Hallidays had to concern themselves with all other aspects of running a sheepfarm. They explored every inch of their land. They gave names to hills they had not yet christened: Bush Hill and White Hill to the west, Bell Hill and High Hill to the north and the Sugar Loaf out on the Atlantic coast. They located the exact whereabouts of freshwater springs, the largest, behind Square Hill, soon to become known as "Gransy's Springs", as it was Grandpa McCall's favourite spot at which to sit and smoke his pipe. A mile and a half to the north was a saltwater lake where, judging from the number of arrowheads and flint chippings found, Indians had at one time or another kept their tents. They established which gullies or cañadones offered best shelter for the sheep, how the pastures varied hectare by hectare and how many head each paddock could be expected to support, this depending on the density of vegetation, the quality of surface soil and the proximity of fresh water.

Everywhere they looked, they foresaw the tasks ahead. A farmhouse, sheep-pens and shearing-shed had to be built. Paddocks had to be fenced off, a tricky and costly business as merely the perimeter of the land would require some thirty miles of fencing. A campaign had to be fought not only against the pumas but against the rheas and shell-geese, both competing with the sheep for pasture. And there was no knowing how the sheep purchased from San Gregorio, themselves imported from the Falkland Islands, would adapt to the conditions of Hill Station. For the time being, however, it was a question of "Beggars cannae be choosers".

With one horse, six sheepdogs and no outside help, the running of the farm was a family affair. William was, of course, "Skipper". A serious-looking man, with a "sleeping" left eye, he left most of the smiling to the other members of the family, but no one ever recalled him once losing his temper. Mary, too, was a quiet person with the happy and vital knack of making light of difficult times and of making a meal out of next-to-nothing. She seldom complained and it came as a surprise to strangers when they first noticed she had two fingers missing from her right hand, the result of a riding accident in the Falkland Islands. She was going to have to be tough and stalwart.

Grandpa McCall was what might be described as foreman or *capataz* of the farm, making sure that things got done. He had also suffered a riding accident in the Falkland Islands, breaking his wrist so severely that he was unable to ride and was severely restricted in the amount of manual work he could do. An imposing man with bushy white side-whiskers, like his son-in-law he stood at six feet two inches, "as tall as any of those giants". And his eyes were as flashing as his temper. His grandchildren took it in turns to syringe his wrist each morning, and they knew that it was wise to keep "in Gransy's good books" as he imposed on the farm what he would have considered an essential, kind discipline.

Having been brought up in a part of the world that lacked anything resembling normal entertainment, the Halliday children were in many ways mature for their age. Great responsibility fell on the shoulders of Willie, the eldest son. Only eight years old, he was already a proficient horseman and carpenter. His sisters conceded that "he knitted the best pair of socks of us all". Agnes Jane, the eldest child, known as Mack, was considered a second mother by her brothers and sisters, even though she was still only ten. She helped with the general running of the "house", especially during the time that Mary was occupied with keeping baby Archie alive. In contrast, the second daughter, Mary, was already something of a tomboy at the age of nine. An excellent horsewoman and "good with her fingers", she was given the nickname "Mendie" because, preferring the life of a shepherd to that of a cook, she was for ever mending things about the farm. The three other children, John, aged six, Margaret four, Annie three, and baby Archie, were too young yet to play their full part in the running of the farm, but they would be expected to take their turn in the years ahead.

Each season would bring problems of its own, but these spring months were the busiest of the year. The sheep were thin and weary after winter, especially as a result of the journey from San Gregorio, and the "shepherd of the day" would often have to dismount to skin a dead sheep — if carancho and condor had not already obliged — or to lift up an animal that had fallen on its back and was unable to move because of heavy pregnancy or the sheer weight of its wool. Sheep had to be dug out of the drifts caused by sudden snowstorms, and there was the ever present threat of a puma, and the rounding up of the flocks into the corrals each dusk.

The first event in the sheepfarming calendar, lambing, took place in November. The family were already well acquainted with the procedure: the recently-born lambs were separated from their mothers, their tails lopped off, and if there were too many tup or male lambs they were castrated. At the end of the day, the tails were counted, and, for the less squeamish members of the family, the testicles were grilled on the open fire. That first year, 1885, the Hallidays counted about 250 tails, representing an increase in the flock of approximately forty per cent. In a successful year, they could expect a growth of up to eighty per cent, but they also knew that, after a bad winter, there might be no increase at all, the ewes becoming so weak that they were incapable of giving birth or supporting a lamb even if one was born. In the circumstances, forty per cent was "nae bad".

The following January, the first shearing took place. Willie singled out the sheep to be shorn, and William, with a single pair of clippers that was another item of Grandpa McCall's godsend box of tools, shore all 700 sheep by hand. Agnes Jane and Mary carried the fleeces to a makeshift table, twisted them tight and "jumped" them into boxes made from driftwood, using a crowbar to pack in as many as possible. Grandpa McCall sat nearby and gave encouragement and tea. The "bales" were then rolled down the small cliffs or *barrancas* to the beach to await the arrival from Punta Arenas of one of Bráun y Blanchard's wool-ships, God knew when.

It was at this time that the Tehuelche Indians at last made contact. The Hallidays watched with fear yet relief as an Indian and his squaw rode past Square Hill towards the tents, fear that they had come to cause trouble or to reconnoitre the flocks and valuable bales, relief at finally meeting a people about whom they had heard such conflicting reports. The information William had been given about the Patagonian Indians was as diverse and uncertain as that about Patagonia itself. They were on the one hand ruthless fighters, on the other hand peaceful towards settlers, pilferers and scroungers yet hospitable hosts, unreliable labourers yet brilliant horsemen, gullible yet self-confident, dissolute and drunk yet dignified and strong. The problem was that the Indians had suffered as much from generalization as the country in which they lived. Just as Darwin's curse of sterility had been taken by some people to apply to the whole area, so the experiences of early Spanish colonists south of Buenos Aires gave all Patagonian Indians a reputation for being primitive and bloodthirsty. Many experienced seamen preferred not to land at all, any trading being carried out in a boat offshore. The tribes were in fact many and varied, some sedentary and warlike, others nomadic and peaceful, some as different from each other as the cordillera of the Andes from the open pampa. The Yàmanas or Yaghans, who lived on islands off Cape Horn, went about almost naked, used small canoes and spoke a soft, rapid language that contained over 30,000 words but whose numerals stopped at five. The Onas, who lived on the main island of Tierra del Fuego, were insular to the point of not using boats, and spoke slowly and in harsh, guttural tones. The western channels of the Straits were inhabited by the Alakalufs. The Manzaneros or "Apple People" lived among orchards planted by seventeenth-century Jesuits south of Lago Nahuel Huapi, the Araucanos north of Nahuel Huapi and beyond, the Pampas north of the Río Chubut, the Northern Tehuelches between the Rios Chubut and Santa Cruz, and the Southern Tehuelches between the Río Santa Cruz and the Straits of Magellan. These tribes were themselves divided into groups and sub-groups. For example, the most northern Araucanos were known as Abajinas, those in the heart of the Andes as Arribanas or Moluches, after their hereditary chief Malechou, those on the Pacific coast as Costinas and those to the south as Huilliches. The Tehuelches, themselves originally a sub-division of the Puelches or "men of the East", were divided into Levuches or "men of the river", Calille-het or "men of the mountains" and Huilliches or "men of the sea". Those with whom Musters spent a year were another branch known as the Aóni-kenk, with their own dialect, Aóni-aish. Different tribes had different names for each other, the Araucanos being known by the Tehuelches as Picunches, the Pampas as Pehuelches and the Manzaneros as the Chenna.

For an average colonist, settler or member of the Argentine and Chilean militia, this diversity was unintelligible and probably unimportant. The Indians of Hill Station should have been Southern Tehuelches, but by the 1880s it was impossible to be sure how many of them were of pure blood. Formerly a nomadic tribe, they were known to have travelled as far north as the hills of Tandil and even to the plains of Buenos Aires itself. Towards the end of the eighteenth century, their great enemies the Araucanos, whom they had managed to keep at bay for aeons due to their superior knowledge of the terrain, surprised them at what became known as the Pampa of Languiñeo, or the Place of Many Dead. Large numbers of Tehuelches were captured and forced to live with the enemy, but survivors fled southwards and planned frequent attacks on Araucanian strongholds. Facing the threat of extinction and the ruthless opposition of national forces, however, tribes were obliged to discard old feuds and to join forces for the sake of common survival, resulting in much intermarriage and the sharing of encampments. Although the Patagonians themselves preferred to be acknowledged by outsiders as *paisanos* or countryfolk, they were almost invariably referred to simply as "Indians", they themselves calling all outsiders *cristianos*.

Of whatever blood, the Indians at Hill Station were the cause of some disappointment to the Hallidays on this and subsequent occasions. Were these, then, the descendants of the infamous "Patagons"? Were these the offspring of the giants? They were reasonably tall and commanding, with high cheekbones, large hooked noses, arrogant and weather-beaten faces and remarkably white teeth. The broad shoulders and muscular arms of the man gave some hint of how the Patagonians had acquired their reputation for magnificent acts of strength. The only clothing they wore was a guanaco mantle and loincloth. When the Hallidays later asked how they could endure the wind, rain and snow with such meagre protection, they replied: "You see, we are all face!"

But they looked dissipated, and they stank. The Scotsmen, who themselves had been living in squalid conditions, did not react as strongly as Lady Florence Dixie at her first meeting with a Tehuelche: "Whatever he may have thought of us, we thought him a singularly unprepossessing object, and, for the sake of his race, we hoped an unfavourable specimen of it." But they felt something of an anticlimax, like the traveller in Bolivia and Peru who has been led to believe that the Indians of the high altiplanos are descendants of the Incas, or like the naturalist who first sees the condor within the dismal confines of a zoo.

The fact was that the Hallidays were to meet only the residue of a race. The number of Indians throughout Patagonia had fallen drastically in recent years. In the 1830s Darwin estimated that there were still 10,000 members of the Alakaluf tribe alone, but by the 1870s Musters put the entire Indian population of southern Patagonia at under 3,000, "of which perhaps 500 were fully-fit men". When Orkeke, last of the great Tehuelche chiefs, was captured and taken to Buenos Aires in 1883, his tribe consisted of seventeen men and thirty-seven women and children. When the Hallidays arrived, there were probably no more than 300 Indians surviving in the Río Gallegos area. It was to be a sad irony that the first cristiano settlers should on the one hand live in harmony with the indigenous people and on the other hand witness their extermination. Predictions made by earlier visitors, such as Beerbohm and Hudson, would prove all too true: the Indian "would survive only in memory as a sad illustration of the remorseless law of the non-survival of the non-fitted"; "some of their wild blood will continue to flow in the veins of those who have taken their place, but as a race they will be blotted out from earth, as utterly extinct in a few decades as the mound-makers of the Mississippi Valley and the races that built the forest-grown cities of Yucatan and Central America". In their diaries and reminiscences, the Hallidays spoke as much of a more glorious past as of the contemporary scene, as much of the way in which the Indians slowly but inexorably disappeared as of the part they played in the development of Hill Station and southern Patagonia.

On this first occasion the Indians made it clear by gestures that they had ridden to Hill Station for the peaceful purpose of trade. The squaw extricated a quantity of ostrich-feathers and guanaco-mantles from the saddlebags of the packhorse. William explained by similar means that he was not interested in such things but wished to purchase a horse. The haggling had begun, not that the Hallidays had much with which to barter: a little of the ship's biscuit, perhaps some of the less-needed tools, or a pinch of Grandpa McCall's tobacco or, just perhaps, the bottle of Scotch whisky which he kept carefully hidden, "purely for medicinal purposes" as his wrist could at times be very painful. William must have been reluctant to offer a sheep as it was not in his interests that the Indians became partial to the taste of mutton.

Business transactions between Indian and cristiano were frequent in Patagonia, and it was to the latter's gain that the other did not possess the same set of commercial values. Falkner had seen a fox-skin poncho, "as fine and as beautiful as ermine, worth from five to seven dollars", willingly exchanged for a string of beads "worth about four pence". Piedrabuena bartered a small sack of maté for an Indian horse, Akers a pair of lace-up riding boots for a mare, the Welsh a loaf of bread for half the carcass of a guanaco, and, when the Hallidays at last acquired some flour, their scones and pancakes would be considered luxuries.

In bartering, a cristiano often needed what Beerbohm called "the patience of Job and the temper of an angel". Apart from the delight the Indians took in haggling and in prolonging any social occasion, there was the obvious difficulty of language. Some Indians spoke a little Spanish and had picked up a few words of "Anglish" from sailors and missionaries, and from Musters, but they were unlikely to understand anything said in a Dumfries brogue. Likewise, it would be several years before the Hallidays learned Spanish or the basic words of the Indians' language, Tsoneca.

The Indians finally made it clear that they would be willing to exchange a horse for their favourite commodities, tobacco and alcohol. William regretted that now, and on subsequent occasions, he was obliged to deal with the Indians in liquor. Like the Welshmen in Chubut, he soon learned that it represented an irresistible source of pleasure and was by far the most valuable item of barter in the cristiano's cupboard. A few Patagonian tribes made alcoholic beverages of their own. The Manzaneros brewed cider by fermenting apples in pits lined with horse-hide, and each autumn had riotous drinking-bouts. But most were introduced to rum, brandy and caña by European sailors from the sixteenth century onwards, with devastating effect. By the eighteenth century, according to Falkner, alcohol was wreaking havoc among the tribes of the north. By the mid-nineteenth century Akers wrote that it was having a degenerative effect throughout Patagonia: "They are one and all totally incapable of resisting the fascination of caña if it is by any means obtainable, and when any kind of intoxicating liquor is to be had they will remain in a hopelessly drunk state for weeks on end." Tehuelches were known to pawn or sell their wives and children for the sake of alcohol, despite the importance of their attachment to family life. When a child, the cacique Casimiro had been "sold" to the governor of the Río Negro settlements for a cask of rum and, perhaps as a result, later succumbed to the addiction himself, exchanging all his possessions except for two horses for himself and

Hasta los viejos indios conocían las excelentes cualidades de la GINEBRA "BOLS"

his family, with hardly any riding-gear between them. Piedrabuena and Clarke tried to help him but it eventually became hopeless to give him anything as he immediately exchanged it for alcohol.

It might have been of comfort to those who objected to this form of trade that, at least at times, the Indians took precautions before a drinking session. Among the Tehuelches it was usually agreed that one male should remain sober to protect the families, and among the Manzaneros, according to Hudson, "the women would first go round carefully gathering up all knives, spears, bolas or other weapons dangerous in the hands of drunken men, to carry them away into the forest, where they would conceal themselves with the children. Then for days the warriors would give themselves up to the joys of intoxication". The Indians were, Musters wrote, as prone to hangovers as any heavy drinkers: "The party went out hunting in the morning, the ride no doubt proving beneficial to those suffering from headaches."

Many cristianos were quick to exploit the Indians' susceptibility to alcohol. Both the Argentine and Chilean authorities included caña in official rations distributed at Punta Arenas and Isla Pavón, thereby encouraging the Indians' natural apathy and helping them towards what may have been considered as their convenient extinction. Traders turned the weakness to their advantage. Prichard wrote that one "of Paraguayan extraction, with a full greasy-lipped animalism" made a point of visiting the tents or *toldos* at times of festivals and tribal ceremonies, when the Indians were almost invariably drunk. Another organized horseraces on the outskirts of Punta Arenas and then challenged the winner, usually taking both the race and the bet because of the drunken state of the Indian rider. Lady Florence Dixie deplored the fact that the Indians, seldom sober during their visits to Punta Arenas, "generally get worsted by the cunning white man".

Before handing over his side of the bargain, William decided to accompany the Indians to their toldos both to select his horse for himself and to take a closer look at the people who were to be his neighbours. As had been expected, they were encamped by the inland lake, a place they called Kippern Aike or Good Hunting Ground. An "aike", common in place-names throughout southern Patagonia, was the equivalent of the Spanish *paradero*, a stopping-place at which the three essential ingredients for survival were to be found: meat, water and firewood. A few hundred yards above the lake were some extensive freshwater springs, and the Hallidays were later to call the place Los Pozos, Spanish for The Wells.

William had seen toldos from a distance on his exploratory trip to Santa Cruz but this was the first time he had been at close quarters and, as he approached, he heard the barking and yelping of countless dogs. Each toldo contained up to thirty animals, described by Musters as "so mixed in race as to defy specification", and by Hudson as "a generalized beast, grandson of the jackal and first cousin to the cur of Europe and the eastern Pariah". Most were hunters but some, generally smaller and noisier, were purely pets. Musters complained: "These little lap-dogs are the torment of one's life in camp; at the least sound they rush out yelping, and set all the bigger dogs off; and in an Indian encampment at night, when there is anything stirring, a continual concert of bow-wows is kept up." Seeing the dogs, a visitor might have wisely followed the advice of Martín Fierro: "Don't trouble to halt at any place where you see lean dogs around." Despite their emaciated appearance, the hunting dogs were treated with indifference but seldom with cruelty. The Hallidays saw nothing to confirm Coan's claim that "their owners were almost constantly beating them with clubs, keeping up an incessant roar in the canine family". On the contrary, if a Tehuelche couple had no children, they were likely to lavish affection on their lap-dogs instead. Musters recalled that when the tribe of the childless chief Orkeke was once crossing a turbulent river, the women and children were sent down-stream to the safety of a ford and were accompanied by one man with specific orders "not to allow his little dog to get wet".

Fortunately for a visitor, the bark of a Tehuelche dog was usually worse than its bite. If anyone tried to enter a toldo unaccompanied by its owner, however, he was likely to regret it. Musters wrote of one such incident:

A young gallant had sought admission to the toldo of his inamorata by the accustomed method of cautiously lifting the back tent-cover from the ground, and dextrously crawling underneath; when half through, he felt his legs seized in a pair of powerful jaws. The lady was highly amused at the predicament of her lover who, however, extricated himself by a mighty and well-directed kick with his foot on the muzzle of his assailant. When returning from his rendez-vous he met his active enemy and vindictively knocked him on the head and, to make sure his work, cut his throat; but his leg carried after all a deeper scar than his heart as a token

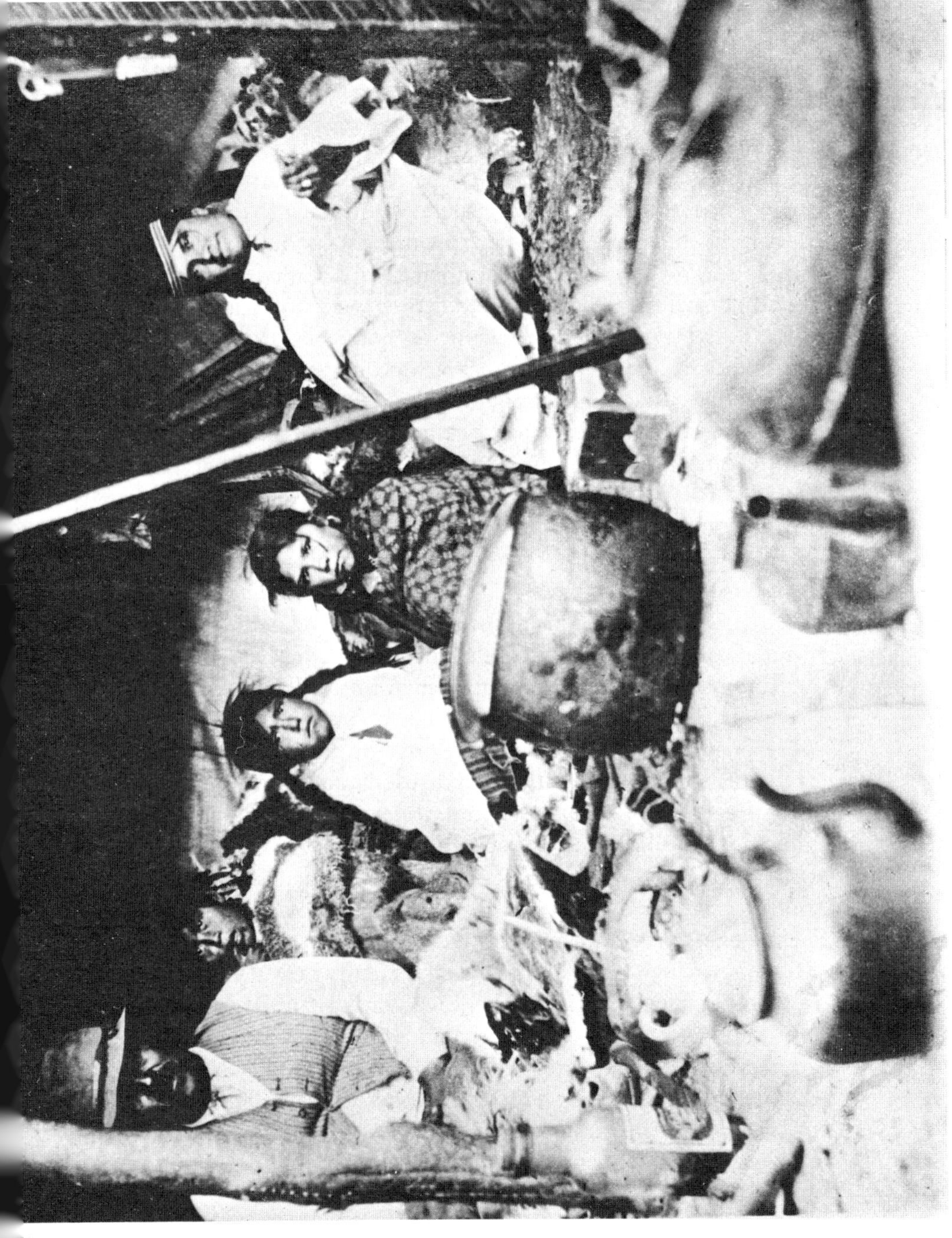

of his love-adventures, and when the story was told and, as may be supposed, excited roars of laughter, it recalled forcibly to mind — "He jests at scars that never felt a wound."

The toldo itself was a ramshackle sight. Prichard wrote: "Imagine a very squat deep-draught boat cut off at rather beyond half of her length and turned upside down." It was erected in the following way: a row of posts about three feet high was driven into the ground and a ridge-pole laid across, another row of posts about six feet high was placed several feet in front, and then a third row of posts about nine feet high. All the poles slanted a little but by no means in the same direction. A canvas made from as many as fifty guanaco-skins smeared with grease and red ochre was dragged over the framework from the rear, its weight straightening at least some of the poles. If the weather was particularly bad or the Indians were settling down for winter, an additional canvas was secured to the front poles and fastened to an extra row of short posts. Relatives and close friends would sometimes combine their toldos by overlapping the canvases so that two or three different domestic interiors were under the same roof.

To a *cristiano* the inside of a toldo was usually repulsive. As Bourne put it: "Shakespeare's imagination never comprehended anything so detestable as a Patagonian hut." The sleeping-quarters were divided from the rest by guanaco-skin curtains. Belongings were piled up round the sides to prevent draughts. The furniture, consisting of bolsters made from disused mantles stuffed with straw and threaded with guanaco sinews, were used as saddles, as pillows at night and, by the women, as cushions during the day. If the ground was particularly damp, the men would spread out their saddle-cloths, but they usually preferred to sit on the bare earth. Some Tehuelches were particular about the cleanliness of their toldo — Musters wrote: "A patch of sod accidentally befouled is at once cut out and thrown outside by the women" — but most were not. Normally everything was riddled with lice or *cherro*, which, according to Coan, were treated as a delicacy: "One with a club beat the fur side of the mantle, while the other watched the game as it was thrown out of the fur, seized and devoured it with great relish." The Hallidays saw no such thing, but what they could not fail to notice was the stench.

In one corner of the toldo would lie a hotchpotch of provisions acquired from a passing trader or on the latest visit to Punta Arenas or Puerto Santa Cruz: coffee, maté, sugar, flour, biscuits, pasta or *fideos*, black beans, maize, and so on. Any tobacco or alcohol was probably to be well concealed elsewhere in the toldo. There was likely to be an ox-horn tinderbox and a leather bucket for fetching water. If a hunt had recently taken place, the carcass of a guanaco or rhea would be hanging up outside the toldo, out of reach of the dogs which in turn would frighten off the foxes, condors and caranchos. A supply of dried meat or pemmican was kept in reserve for the emergency of a hard winter but the Tehuelches generally preferred to eat their meat fresh and raw. The lungs, heart, liver, kidneys, even the most offensive parts of the intestines, were consumed unwashed and uncooked, with perhaps a touch of salt or ground pimento, as was the fat from under the guanaco's eyes and from behind its thigh joints. The greatest luxury was fresh blood. On special occasions, a young mare was trussed up and suspended by its legs, its throat was cut and the blood was collected in bowls and swallowed ceremoniously, still warm. This habit shocked most cristianos and, for some at least, typified the true nature of the Indian. In the words of Hernández:

> The only thing in his savage creed
> That the Indian's sure about
> Is this; that it's always good to kill,
> And of smoking blood to drink his fill;
> And the blood he can't drink when his belly's full
> He likes to see bubble out.

In recent years, however, the Indians' manners had been influenced by contact with cristianos and by an increasing dependence on government rations. They now often cooked their meat. Besides a large iron pot in which to extract ostrich grease and marrow, each toldo possessed a kettle and a grill for cooking meat and pancakes, or *tortas fritas*, flour having become a much prized commodity.

It was not so much because of squalor, lice, foul smells and grim eating habits that William Halliday forbade his children ever to enter an Indian toldo and only ventured inside himself when, as on this first occasion, he was inquisitive and did not wish to offend hospitality. It was primarily because of disease. This was the Indians' most feared enemy, against which neither the indigenous remedies nor the cristianos' medicine were of any avail. Whether it attacked in the form of tuberculosis, typhoid, measles, whooping-cough, diphtheria or smallpox, it met with little resistance. The Hallidays found it difficult to understand how these sturdily built Patagonians could succumb so easily even to the common cold. The truth was, once again, the catastrophe of the Old World introducing its diseases to the New. In the eighteenth century Falkner was already recording the fate of the Levuches, who once roamed the banks of the Río Negro:

> After an attack on the plains of Buenos Aires, they retreated to the Vulcan but unfortunately they carried away with them some clothes which a short time before had been bought in Buenos Aires and were tainted with smallpox. There were now only 300 fighting men left.

The "numerous" nation of Chechehets, too, had contracted smallpox near Buenos Aires and made a futile attempt to escape from it by retreating to their own country "about 200 leagues distant". By the late nineteenth century, disease was widespread among all Patagonian tribes. George Catlin, the North American ethnologist and artist, once noticed the scars of smallpox on the face of a Tehuelche cacique near Punta Arenas and asked how the disease had been contracted: "We are poor, we want many things that the white people make — their clothes, their knives, their guns and many other things — and we come here to buy them. . . . We do all we can to prevent this but it is still not stopped, and we are afraid of getting the awful disease again." Another group of Tehuelches had picked up contaminated waste-paper from Puerto Hambre from which they contracted smallpox and, by association with the disease, a strong suspicion of the written word.

If unfortunate enough to fall sick on the march, an Indian was left where he was, as Falkner put it, "forsaken and alone, with no other assistance than a hide reared up against the wind and a pitcher of water". If taken ill at an encampment, his toldo was separated from the others, and, if he died, it was at once burned to the ground.

Each cacique formerly had two men responsible for the medical treatment of those under his care, the *medico* or doctor and the *brujo* or wizard. The former dealt with external injuries and was a member of the tribe who had acquired a reputation for mending broken bones. The latter dealt with illness as opposed to injury. The Tehuelches believed that all disease was a bewitchment cast on an individual either by the evil spirit Gualichu, or by the cristianos, or by both.

The brujo had once held great power. The position was usually granted to a female, and any male incumbent was expected to dress like a woman, preference being given, according to Falkner, to those who at an early age discovered "an effeminate disposition". At the outbreak of an epidemic, the cacique and all his tribe visited the brujo's toldo, always kept well apart from the others. Ceremonies took place which, certainly through the eyes of a Jesuit priest, seemed strange and primitive:

> They assemble in the wizard's tent, who is shut up from the sight of the rest in a corner of the tent. He has a small drum, one or two rounded calabashes with small sea-shells in them, and some square bags of painted hide in which he keeps his spells. He begins the ceremony by making a strange noise with the devil, who it is then supposed has entered into him; he keeps his eyes lifted up, distorting the features of his face, foams at the mouth, screws up his joints and, after many violent and distorting motions, remains stiff and motionless, resembling a man seized with epilepsy. After some time, he comes to himself, as having got the better of the demon; he next feigns, within his tabernacle, a faint, shrill, mournful voice, as of the evil spirit, who, by this dismal cry, is supposed to acknowledge himself subdued; and then, from a kind of tripod, answers all questions that are put to him.

Whatever power this gave the brujo, she or he took full blame for any calamitous consequences. If unsuccessful in bringing an epidemic to an end, the brujo would try to convince the tribe that the bewitching had been caused by a particular individual:

> In a muddy pool they drowned the boy
> As the cause of the plague of pox.

If a cacique died, the brujo usually died with him, especially if there had been a recent argument between them. But, as the Indians sunk deeper into apathy, the brujo's power dwindled. At Kippern Aike she still kept her toldo well apart from the others, but on the few occasions the Hallidays managed to catch sight of her she was as drunk as most of her protégés. Even if she had lost all influence in communal decisions, however, she still received individual visits from the sick and the aggrieved. She was said to possess a magic stone full of holes and had only to place the tiniest piece of clothing or a hair or some saliva inside it for a spell to be cast. A bewitched Indian would then beg or bribe her to remove the spell. If successful, she was rewarded handsomely, if unsuccessful she was best advised to

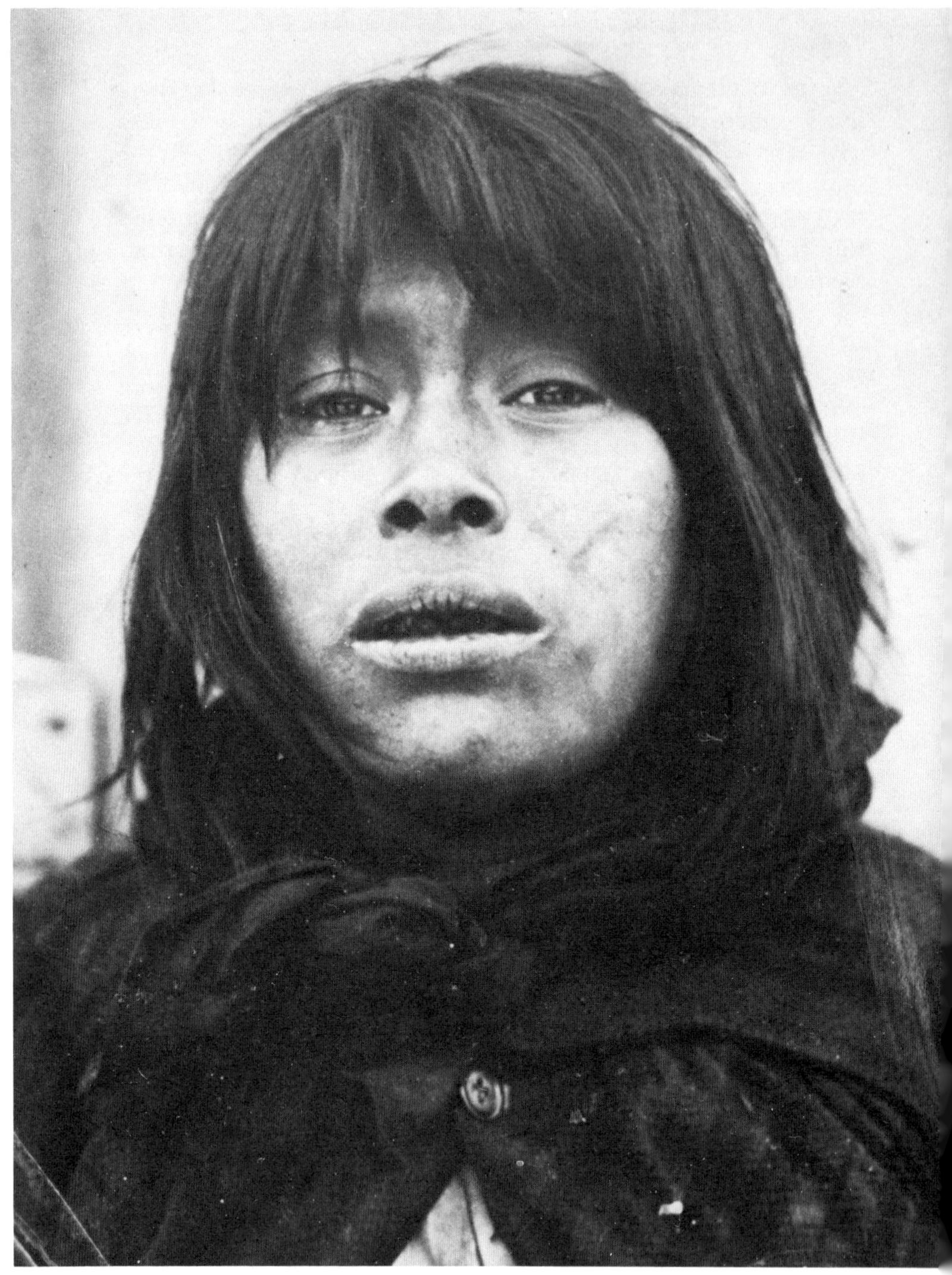

pack up her toldo and leave in the silence of the night.

In the cure of illness, her remedies often seemed bizarre. A victim of measles was thought to have been invaded by a Gualichu of such power and ferocity that blotches and spots appeared on the skin. As the Gualichus were known to dislike noise and cold, the patient was made to ride naked through the wind and snow, screaming loudly. A victim of smallpox, on the other hand, was likely to be greased from head to foot and put out to boil in the sun. When Bourne suffered from lice, the brujo sucked blood from his neck, growling all the time. In other cases, the brujo would sit on the patient, wail, suck the thumbs, strike the breast, blow through the fists, lick the eyebrows and face, and breathe heavily over the whole body. Henry Jamieson once heard the groans of a young girl and found her mother "filling her daughter's mouth with mud to keep the devils out".

The decline in the influence of the brujo reflected the fact that by the late nineteenth century the Indians' beliefs in general had been clouded by alcohol and by an alien culture half-absorbed. This was particularly true in their attitude to death. It had traditionally been held that the spirits of the dead were carried to heaven on horseback, that they were there transformed into stars, that the Milky Way was the field where they hunted, and the Magellanic Clouds (two circular cloudspots consisting of vast numbers of nebulae and clusters of stars only visible in the southern hemisphere) were the feathers of the ostriches they killed.

A funeral had formerly been a complicated affair. As soon as someone died, the body was forced into a sitting position by a strong woman pressing on the shoulders and snapping the spine. Still dressed, it was placed at the entrance of the toldo. At sunset the women started to dance mournfully round the tent, their faces blackened with soot. Others consoled the widow or widows. Falkner observed: "During these visits of condolence, they cry, howl and sing in the most dismal manner, straining out tears and pricking their arms and thighs with sharp thorns to make them bleed." The men sat silently and did not weep. All the deceased's animals, except for their favourite horse, were slaughtered in the belief that they would suffer too much without the company of their master or mistress. At the same time, all save the most treasured possessions were placed in a pile and burnt to cinders, relatives and friends throwing gifts on the fire. The Tehuelches preferred to be buried close to their ancestors at graveyards along the Atlantic coast and, in days when they roamed more extensively, this could involve a complicated embalming of the body and a journey of several hundred leagues. At Kippern Aike, however, the body was carried on horseback the short distance to a burial-ground north of Cape Fairweather. The widow or widower followed close behind, leading the favourite horse loaded with the few remaining possessions. Wrapped in the hide of a colt, the corpse was placed in its sitting position inside a grave dug by the women, together with a leather-bag full of the ashes from the bonfire, with any article that would not burn, and with salt, sugar, maté and other provisions for the journey to heaven. The favourite horse was strangled to death above the grave and, with its front legs fixed into a kneeling position, was buried next to its owner, the treasured possessions still on its back. The grave was filled, and covered with stones.

In the 1880s, some Indians were still buried in the old manner and the entire contents of the grave were often removed to a secret place to avoid being ransacked by cristianos. The majority, however, now received a simple Christian burial at one of the settlements or farms. Henry Jamieson wrote in his diary of an Indian mother becoming so hysterical at the thought of nails being driven into her daughter's body that "the funeral had to take place with no lid to the coffin; but we also buried the riding tackle and horse to carry the dead girl to their heavens". Likewise, although the rules of mourning had once been strict and a widow was obliged to stay inside her toldo for a whole year after the death, without eating meat or washing her hands or face, in modern times she was more likely to drown her sorrow in hard drinking.

Although several Indians were to become well known to the Hallidays, it was difficult for the settlers ever to ascertain exactly what they and their ancestors believed in. They certainly did not worship idols. Their prayers seem to have been directed almost exclusively at the Evil Spirit, the Great Gualichu, who lived in the cordillera with his retinue of minor Gualichus, such as Setebos, and sent out strange smoke-signals and made ominous sounds, in fact the spume of avalanches and cracking of glaciers. The Indians thought it more logical to plead with Gualichu not to cause harm rather than cajole the Good Spirit, Guayava Cunnee, who was there to do good in any case. Although it was the latter who had created the original Indians in a vast underground cave, providing them with wild animals, lance and bow and arrow, the Great Gualichu was the more powerful, responsible for all evil to man and beast alike, even for the fatigue after a long journey.

Whatever the precise nature of their modern faith, the Tehuelches had remained a superstitious people. They asked blessings from any particularly beautiful place. They saluted the new moon. If a strong wind blew from an unusual direction, they were known to ride out to attack it. According to Musters, if anyone was foolish enough to climb a certain hill that was shaped like a ship, "he would have his mantle blown to pieces by furious gusts of wind in even the calmest weather". A compass was thought to possess magical powers, being extremely popular among card-players, who believed it would bring good luck to whoever was holding it, and when one of Musters's companions asked to be allowed to take it to help restore the use of his arm, "he sat patiently, with an air of awe and faith combined for an hour, afterwards declaring that the operation had done him much good".

It was still easy for a cristiano unwittingly to offend an Indian. As a result of the epidemics caused by contaminated paper, anyone using pen and paper was immediately mistrusted. When nonchalantly placing stones in patterns on the ground, Musters was accused of "mentally composing the death of someone by witchcraft". An Indian with whom the Hallidays were once hunting suddenly dismounted, cut off some of his horse's mane, rolled it into a small ball and carefully buried it among some bushes: "When Archie asked what he was doing, the Indian rode off and would not speak to a Halliday again." A visitor from another tribe could also become an unpopular figure. Casimiro told a story that was typical of those handed down from generation to generation:

> Many years ago, when I was quite young, I was travelling a few leagues to the northward.... The party encamped near a large lagoon not far from the Sengel River, and were occupied in hunting in the neighbourhood. On several days in succession, smoke was observed in different directions, which approached nearer and nearer each time. Being naturally supposed to be caused by the Indians, it was answered, and scouts were at last sent out to ascertain the cause, as no messengers appeared. They returned, however, stating that they could discover nothing. At the end of four days, an Indian, tall, gaunt and emaciated, mounted on a very thin mule, arrived in the camp, and asked for a chief whose name was unknown. The stranger was taken, as is customary, to the chief's toldo, and his mule turned loose; but strange to say, it never

moved from the spot where it was unsaddled and the Indian during the time he remained in the toldo neither ate nor drank. At the end of three days he mounted his mule which appeared as fresh as when he arrived and rode away to the northward. On the following day, whilst hunting, a sickness struck the Indians, some falling dead from their horses, while others, although able to return home, only survived a short time.

Despite their confused faith and the adversities of disease, squalor and drunkenness, the Indians of Kippern Aike left a lasting impression on the Hallidays as a people who still enjoyed life and remained essentially benevolent and affectionate. This was most strikingly reflected in the love, almost reverence, they showed towards their children. If a baby boy was born, a mare was dissected from the neck downwards, the entrails torn out and the child placed inside the cavity, thus ensuring that he would be a fine horseman in the future. If a child was injured or merely cut a finger, a feast was likely to take place, in celebration if he recovered, in commiseration if he did not. At the death of an only child, the parents would burn all their possessions, kill all their horses and reduce themselves to a state of abject poverty in their grief. After a hunt, it was the children who were given the titbits.

To a cristiano, this affection could seem excessive. Falkner had complained: "They breed up their children in a vicious indulgence of their humours." He claimed that the old people were led about from one place to another, frequently changing their habitation to humour the caprices of their children. He wrote of the widow of a Tehuelche cacique who had been "trecherously killed by the Spaniards in time of peace" and was determined to leave the area. But her six-year-old son was very fond of the missionaries of Tandil and particularly partial to the gifts they gave him of bread, figs, raisins and so forth. When he realized his mother was intending to move on, the child made a great fuss, refused to get dressed for the journey and insisted on being taken to the priests. The mother was so moved by her son's distress that she not only agreed to remain at the mission, but soon afterwards was converted to Christianity. Grandpa McCall also considered the Indian children to be "downright spoilt", but he and the rest of the family indirectly benefited from this: a reason that the Indians treated Hill Station with respect was that the place was full of children.

Tehuelche women were expected to do virtually all the manual work in the encampment. Besides looking after the children and the

cooking, they fetched wood and water, skinned animals, sewed up hides, wove blankets, packed and unpacked provisions, erected and dismantled toldos, loaded and unloaded the horses, tightened girths, and buried the dead. In days of warfare, they had carried the lance for their husbands. They were not excused from work by sickness or pregnancy, and if a husband lifted a finger to help, he was liable to be treated with contempt by the other males. Cristianos often found this imbalance unjust. Bourne said of the men: "Exertion of body or mind is their greatest dread." Lady Florence Dixie expostulated: "It is only the men who are cursed or blessed with this indolent spirit. The women are indefatigably industrious." On that first visit to Hill Station, and on other occasions, it was the squaw who led the packhorse and took care of any physical aspect of the transaction. Her husband sat on his horse and grunted commands.

It would be unfair, however, to suggest that the Tehuelche male was always bone-idle. In the days when the Indians were frequently on the march and were involved in inter-tribal feuds and in the wars against the Argentine militia, the men were collectively responsible for the protection of the tribe and individually responsible for providing their family or families with clothing, shelter and food. In order to survive, they had depended less on receiving crumbs from the cristianos' table and more on the almost full-time hunting of wild game.

One man who could rarely be accused of indolence was the cacique. By Tehuelche law, every individual had to be under the wing of one or other of the chiefs, whose obligation it was to settle any differences and to pass judgment. His word was usually final but, as in all societies, there were those people who preferred to take the law into their own hands. Falkner commented: "When the offence is not very great and the offender poor, the party injured generally beats him with his stone bowls on the back and ribs. When the offender is too powerful, they let him alone, unless the cacique interferes and obliges him to make satisfaction." The cacique decided when and where to pitch camp, when to set out on the march and which route to take. He was expected to be an established orator, frequently summoning the tribesmen to his tent and haranguing them for up to an hour upon their behaviour, the exigencies of the time, the injuries they had received and the measures to be taken: "In these harangues, he always extols his own prowess and personal merit." On the other hand, the cacique had no power to take anything away from his

subjects or to make them work without pay. He was expected to treat them humanely and generously or else they would seek the protection of another cacique. The position was hereditary, but because of the responsibilities involved it was not unusual for a cacique to abdicate in favour of his son or closest relative.

However chauvinistic or indolent in his attitude towards women in general, a Tehuelche man usually treated his wife, or potential wife, with great affection. He was expected to lavish gifts on her, such as brass earrings, beads and, in more modern time, liquor. Courtship was at first restricted to the eyes. Groups of neighbouring Indians were sometimes seen riding towards Kippern Aike "presumably to reconnoitre the womenfolk". When convinced that a girl was not indifferent to his attentions, a young man asked several of his close friends to hold a discussion with the girl's father, at which they would extol the young man's abilities, possessions and prospects. The father then discussed with the girl's mother the terms on which they would be willing to give their daughter away. Typical demands were a horse for each of the girl's male relatives, a mare for her mother to ride and a wild mare for each female relative to consume. If the girl was very pretty, the price would include skins, blankets and silver, especially decorated spurs, bridles and reins. Once informed of the terms, the prospective bridegroom had ten to fifteen days in which to pay, his friends silently leaving the animals outside the relatives' toldos, he himself handing over the more precious gifts. When total payment had been made, the girl was placed inside a special chamber in her father's toldo. The young man quietly joined her in the middle of the night and, if the two were still together at dawn, the marriage was official.

Unless an orphan or widow, the girl had little choice but to accept her husband once the sale was completed. If the marriage was not a success, however, the man simply changed wives. Casimiro was typical in having married at least six times. His reasons, too, may have been typical, according to Musters:

> Certainly, if all his wives were of the appearance and disposition of his last venture, it is not to be wondered at if he disposed of his former ones, for an uglier, more contumacious old hag never burdened the earth with her weight, owing probably to which latter quality or quantity, she never, if she could possibly help it, quitted her room.

Separation was not always, however, the result of a husband's displeasure. A Tehuelche wife usually stood for no nonsense. During

43

his year with the Indians, Musters only witnessed two matrimonial squabbles, and in both a quick reconciliation took place. One rare argument perhaps gives some clue as to why there were not more: "It began by Tankelow's striking his daughter, which his wife angrily resented; from words they came to blows, and the squaw was getting rather the best of it when Mrs Orkeke interposed with a strong arm, and forcibly put a stop to the disturbance."

A Tehuelche wife was as capable as any of slandering her husband behind his back. One of Casimiro's wives, believing that he had deserted the main party of his tribe for fear of being killed, shouted out in public that he possessed "the heart of a skunk, a vulture and an armadillo". And a wife was not always completely faithful to her husband, as Falkner observed: "The contumacy of the woman sometimes tires out the patience of the man, who then turns her away, or sells her to the person on whom she had fixed her attentions. But he seldom beats her or tortures her." Beerbohm was once approached by an old squaw "who was painfully of this world and the vanities thereof, and who had previously been addressing me for some time in the most eloquent of tones". Finding that nothing could melt his icy demeanour, the squaw suddenly hugged him in a greasy embrace. Beerbohm had to leave the camp hurriedly, probably agreeing with his half-brother, Max, that "Most women are not so young as they are painted". However, he called at the toldos the next morning to pay his "visite de digestion" and made a point of going to the old squaw's tent.

> She had a bad headache, and looked very demure and penitent. I rallied her with smiles and attempted, by my assiduous attentions, to make up for my discourteous behaviour of the previous evening. But she took no notice of me and received all my overtures with the utmost indifference. It was a case of "the devil was sick, the devil a saint would be".

Musters was once roused from his sleep "by a lady from a neighbouring toldo who wished to embrace me, and, with feminine curiosity, wanted to know the contents of my letters. She was, I am sorry to say, in an advanced state of intoxication, so, after giving her a smoke, Orkeke, who had roused up and was dying of laughter, politely showed her the door."

It was not uncommon for cristianos, especially "drifters" and peons, to procure a Tehuelche wife or mistress. Jimmy Doig, an Australian who later kept a photographer's shop in Puerto Coyle, was known to have fathered several Indian children, and Governor Ramon Lista, who took over from Governor Moyano, had a half-Indian daughter called Catalina. Not that such relationships were always a success. Isidoro, a deserter from the Argentine militia, confessed to Beerbohm that his days with a girl from the tribe with which he had escaped were not entirely happy:

> Mrs Isidoro, it appears, took to drinking, and became too noisy and violent for her husband, who of all things loved quiet; so, without any further fuss, and without many words, as was his custom, he led her back to her father's tent, where with a short explanation he left her, thus consummating his divorce "a mense et thoro" with expeditious ease, and securing for himself the blessing of undisturbed peace for the future.

Musters and Beerbohm at least contemplated marrying a Tehuelche girl, if only out of curiosity, but neither was able to agree terms with his prospective parents-in-law, Beerbohm offering eight mares, a bag of biscuit and some sugar in return for a girl and five guanaco mantles, but her parents agreeing only to four mantles.

Nothing would have been further from William Halliday's mind than the idea of one of his daughters marrying a Tehuelche, but he would probably have conceded, with Lady Florence Dixie, that "the women can by no means complain of want of devotion to them on part of the men".

This generally affectionate attitude to child and wife was part of a philosophy that Life, however short and diseased, was for living. The Hallidays remembered the Tehuelches as a people who often laughed. They liked to paint their faces, either with a kind of red earth, or with charcoal, or with a pigment made of clay, blood and grease. Unlike some primitive peoples, they enjoyed being photographed. Music and dancing were popular, and plenty of excuses were found for holding a party: to celebrate a birth, a successful hunt or recovery from a minor injury. Even if an old widow bewailed the passing of the good old days, an impromptu festival might be held for her benefit. Normally great trouble was spent on preparations and, by the sound of it, a party at Kippern Aike could be as riotous as hogmanay. The women first emptied the chosen toldo and beat the floor and sides in order to drive off any Gualichu that might be lurking inside. The men meanwhile mounted their horses and, armed with their boleadoras and making as much noise as possible, chased the evil spirits out of the vicinity. The toldo was then decorated with bolsters and ponchos, blue and red being the favourite colours, and a large carpet was placed in a corner where the owner and the cacique would sit throughout the festivities. The owner formally invited his guests to partake of the ribs of a mare and other delicacies, eaten either cooked or raw in front of a large bonfire opposite the entrance to the toldo, the food being washed down with mare's blood, maté or alcohol. Musicians took up their positions next to the bonfire, the orchestra consisting of at least one drum, probably several, made by stretching a piece of guanaco hide across a clay bowl and fastening it with rhea sinews, and an instrument fashioned from the thighbone of a guanaco and bored with holes, played either as a wind instrument or with a short bow across a horse-hair string. Most cristianos could not appreciate the music, Musters writing dispassionately of a song consisting exclusively of the words "Ah ge lay loo, Ah ge lay loo" expressed in various keys, Beerbohm, more accustomed to the music-halls of London, complaining of "the excruciating music of various drums and tom-toms". And the Hallidays continued to prefer the folksongs of Scotland, sung by Grandpa McCall "in a gruff but melodious way".

After everyone had eaten their fill, the women sat in a semi-circle round the fire. The dancing, at least at the beginning of the evening, was a male prerogative and, after cavorting among the women for a while, the men disappeared into the toldo, where the host handed out a number of dancing-costumes, consisting of two straps of ostrich-feathers to be placed over the shoulders and a belt of bells around the waist. The men, camouflaged by paint and ostrich-grease so as not to be recognized by the women, emerged from the toldo and danced several times around the bonfire. At the very last moment before returning into the toldo, they removed their feathers, and the women, recognizing the dancers' identity, began to shriek and sing and make wild gestures, all in time to the beat of the drums. The costumes were handed to another batch of dancers until all the males had had their fling. As the alcohol and music began to take effect, the scene grew wilder, some of the dancers even stepping into the bonfire itself, much to the howling amusement of the women, who themselves gradually joined in with the dancing and were eventually invited into the toldo itself, where everyone stayed until they had fallen into a drunken sleep, or until dawn arrived.

Besides these dances, the Tehuelches organized more sober entertainments. The men delighted in tests of strength, such as lifting above their shoulders heavy rocks like the marble boulder of

Amakaken, which Archie, no weakling himself, was later to try to lift and only managed to raise to his knees. They would sometimes grab each other's hair or neck and try to be the first to force the other to the ground. Or, sitting opposite each other with their legs outstretched and feet touching, they took hold of a stick and tried to pull the other off the ground. They also played a rough form of hockey in which the ball was a knot of wood and the sticks branches of a convenient shape. Much of the time, however, they preferred just to sit at the side of freshwater springs admiring what they called "the eyes of the desert". The children amused themselves in whatever way possible. In winter, they made sledges out of driftwood and were to be seen sliding down the snowbound cañadones. They also made a toy by fastening a pair of rhea feet together. This was thrown in the air and other children would try to hit it — good practice for using the boleadora.

Gambling, like alcohol, was an appendage of cristiano culture to which the Indians fell easy prey. They nearly always had a pack of cards among their possessions, or else they used pieces of guanaco-skin about the size of playing-cards on which dogs, a variety of other beasts and strange scrawls were painted with a stick dipped into a red pigment. Their favourite game was a type of brag, picked up from European sailors and adapted to their own tastes. They fashioned dice from the thigh-bone of a guanaco and also played a variation of knucklebones. They loved laying wagers on all their games and, even when exercising their horses, placed bets on any race that took place. As cash was unknown to them, the stakes were either horses, riding-gear, boleadoras, knives, or black beans, though, as Beerbohm remarked, "as much excitement took place as if each bean or perota had been worth a five-dollar piece". Like true gamblers, they were willing to lose even their "shirt", and a Tehuelche was sometimes to be seen emerging from a cristiano boliche without even his own mantle on his back. One cacique, Hinchel, was typical in having, as Musters reported, "an inveterate fondness for gambling which, together with his lavish good nature, eventually impoverished him".

The Tehuelches had a keen sense of humour. Byron remarked on the delight with which they daubed one of his officers with paint. Musters was once roused in the middle of the night by an Indian troubled with vermin; after sitting for some time, lost in thought, the Indian said, "Musters, lice never sleep." Lady Florence Dixie told of a white trader who was trying to palm off an Indian with a rusty

carbine; after the cartridge had misfired several times, the Indian produced a minuscule piece of ostrich-meat which he offered in exchange for the gun, saying: "Your gun never kill a piece of meat as big as this one. Your gun good to kill dead guanaco." Bourne once fell from his horse into the cold Río Gallegos: "We waded ashore, dripping, amidst the uproarious laughter of the whole troop." And when he showed them his watch they "would stand in every attitude of silent amazement, their eyes dilated, their countenances lighted up in every feature with delighted wonder, and then break out in a roar of hoarse laughter". Beerbohm once made the mistake of losing his temper with an Indian which merely provoked a chorus of mocking laughter among the squaws. And much later, in the 1940s, one of the last pureblood Tehuelches in the Río Gallegos area, a woman called Anna, was to cause some amusement among the Hallidays. Driven into town to have her photograph and fingerprints taken for a national census, she learned that the law was that no spectacles were to be worn, and so insisted that she take off her headband or *vincha*. Unfortunately she also had a little too much to drink while in town and wanted to take some alcohol back to Hill Station. When it was explained to her that this was not allowed as it would encourage the peons to drink, an argument ensued. "Anna suddenly turned and said, 'And you, don't you like little drink sometimes?' To which there was, at the time, no reasonable reply."

The Tehuelches' easy-going attitude to life applied equally to their treatment of strangers. Europeans who fell into their hands, such as escaped convicts from Punta Arenas, explorers and sailors who had jumped ship or were shipwrecked, were usually unharmed. It was, after all, out of the liaison between Indian and *cristiano* that the rumours of an El Dorado in Patagonia had emerged. Whatever else Falkner may have thought of the "Tehuelhets", he found them "courteous, obliging and good-natured". Byron, though more interested in the Indians' height than their manners, recorded that when some of his crew were cutting grass for the sheep on board, the Indians immediately wanted to help, "tearing up all the weeds and trash they could get". A young Irishman, Thomas Craig, sole survivor of a merchantman wrecked in the Straits in the 1840s, fell in with a group of Tehuelches who escorted him safely to Carmen de Patagones. Darwin picked up two men at Puerto Hambre who, having jumped a sealing vessel, had been treated by the Indians "with their usual disinterested hospitality" and, when rash enough to escape from them, "had been living for some time on musselshells and berries, and their tattered clothes had been burnt by sleeping so near the fire". Beerbohm, as ambivalent in his feelings towards the Indian as to Patagonia itself, concluded:

> In general intelligence, gentleness of temper, chastity of conduct and conscientious behaviour in their social and domestic relations, they are immeasurably superior ... all their disadvantages being taken in general consideration, to the general run of civilized white men.

By 1882 the Mulhall brothers, editors of the Buenos Aires *Standard* and authors of the *Handbook of the River Plate*, were able to advise people wanting to ascend the Río Santa Cruz: "The Indians are usually friendly: the traveller had better procure horses and guide from the cacique of the Shehuen Valley." Indeed, some cristianos found the Indians' hospitality excessive. When visiting an encampment near Punta Arenas in 1873, Sir Ralph Williams, later Governor of Newfoundland, was so lavishly treated that he eventually had to swing on to his horse and make a dash for it: "Presumably they meant nothing serious as they did not get on their horses and follow us."

There were, as ever, exceptions to the rule, and some Europeans claimed that they were harshly treated by the Indians of southern Patagonia. The priest Coan, who never took kindly to the place or its peoples, met one William Marshall Thornham, a young English sailor who, having abandoned ship in the Straits, fell in with a tribe who passed knives across his throat and stopped his rations to weaken him. Three other sailors had their clothes taken away, were made to wear "an old greasy and cast-off guanaco skin mantle", and, forcibly kept by the Indians for eight months, were forced to travel on foot and "were often approached with the name servant and slave". In this case, however, the Europeans seem to have received the treatment they deserved: "Instead of becoming chiefs and gratifying their baser passions at will, as expected, they are deprived of every privilege, despised by the savages and degraded to the most abject condition."

Sceptics might have claimed that the Indians' usual hospitality was merely a tactic to give them more time in which to steal as much as possible from a visitor. William had been warned of the Indians' reputation for being inveterate thieves. Perhaps this was a corollary to their increasing apathy and indolence, or the natural reaction of an impoverished people when confronted by the sight of those who have more. It was true that a stranger unlikely ever to return to the toldos was well advised to keep a careful eye on all his belongings. Beerbohm was among those who experienced the Indians' deftness of hand, claiming that "the Artful Dodger would have to look to his laurels if he had to enter into competition with a young Tehuelche". He had learned the hard way. When at an encampment a few miles inland from Puerto Santa Cruz, knowing that English stirrup-leathers were in popular demand because of their buckles he took care not to dismount once during his short stay, but even this precaution proved futile:

> In an unguarded moment, whilst bending over my horse's neck, in conversation with an Indian who was sitting on the ground below me, by way of resting myself a little, I incautiously slipped one foot out of its stirrup and lay at full length along the horse's back. I was hardly two minutes in that position, but it was two minutes too long. On resuming my seat in the saddle my foot sought vainly for the stirrup.

He at once dismounted and went off in vain pursuit of the thief, only to find on his return that the other stirrup-leather had disappeared. With "many a strong interjectional reference" to his own stupidity, he rode stirrupless back to Santa Cruz. Coan also complained that he was often troubled by "the lower class of savages", especially by mothers and their children who would examine his baggage with great care and implore him to open any package to reveal its contents: "They beg relentlessly and pilfer small articles." At Hill Station, the Hallidays were cautious enough to keep a wary eye on Indian visitors who "would sometimes be found lurking suspiciously among the outhouses". Apart from petty incidents, such as the plucking of hair from the tails of range-mares, which fetched a good price on the market, there was to be no case of theft by an Indian recorded in the history of the farm. The Hallidays would come to the conclusion that the reputation for stealing was as ill-found as that of the condor for carrying off livestock. The settlers would have to be more vigilant with the hotchpotch of peons employed over the years, and even with the private tutors later sent out from England.

Some sheepfarmers in Patagonia, however, used the Indians' criminal reputation as an excuse for taking serious reprisals. They murdered them. It was certainly convenient for property-owners and speculators to have the indigenous people removed. The exact

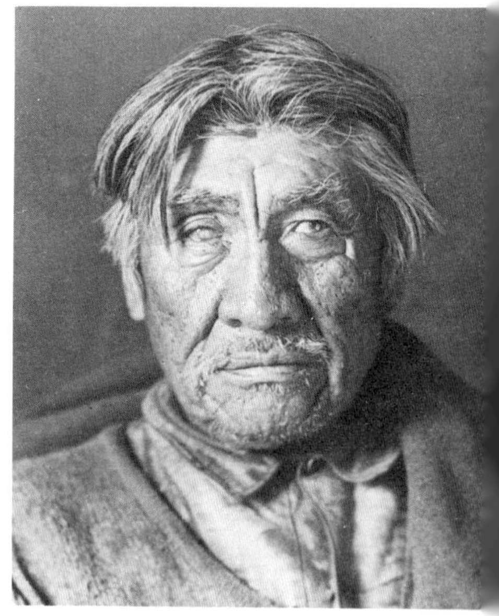

extent to which genocide was practised remains uncertain. Tom Jones wrote from Punta Arenas:

> One or two farmers paid one pound per head to Indian-killers, a few of them British, for those liquidated. Proof of accomplishment was the production of the Indians' ears.

Bill Downer, a Newfoundlander shipwrecked in the Straits who worked for a time on Tierra del Fuego in the late 1880s, later revealed to the Hallidays that he had been given orders to shoot Indians on sight:

> They chopped down the fences with their flint knives, drove the sheep through the gaps to a safe distance and cut the animals' legs to prevent escape.

The issue would erupt in 1928 with the publication of *La Patagonia Trágica,* in which José María Borero accused wealthy landowners, including the Menéndez and Bráun families, of slaughtering Indians both on Tierra del Fuego and in southern Patagonia. The accusations were denied, but rumour bred rumour. The book mysteriously disappeared from shops in Buenos Aires, the whole edition being bought out, it was claimed, by the families accused of the atrocities. The few available copies passed secretly from hand to hand. An intended sequel to the book never appeared.

The Hallidays were of the opinion that some of the claims were probably true. Yet they would also have conceded that, in a way neither intended nor easily perceptible, they themselves were, like all cristianos, indirectly guilty of exterminating the Indians. Any contact helped to weaken their beliefs and way of life, and encouraged their susceptibility to disease and alcohol. By exploiting their gullibility and military weakness, many a trader, government official and member of the militia contributed to this. Missionaries gave European clothes to the Indians who, unaccustomed to wearing shirts and trousers, did not take them off when wet and often fell ill with rheumatism or pneumonia. Most cristianos treated the Indians at best patronizingly, at worst contemptuously, be it in Mulhall's comment in 1869 that a settler on the Río Negro with a good rifle and a well built house could laugh at the Indians, or Lady Florence Dixie's complaint that they rode through her camp careless of her crockery, or Prichard's picnic reference to picturesque heathen camp-servants. The Argentine nation would eventually offer a little of southern Patagonia to the people who had lived their for

centuries. But by then it was too late. The last pureblood Tehuelche died in 1948.

Harassment of the Indians never took place at Hill Station. The Hallidays had neither reason nor desire to kill a people who caused them little trouble and were, in any case, dying. On the contrary, they came to respect and even love the Indians of Kippern Aike, not only for their bonhomie and generosity in the face of adversity but also because they gave advice on survival. Above all, they were of incalculable assistance in providing extra horses and in teaching the use of the boleadora.

Most cristianos conceded that the Tehuelches were the finest horsemen they had ever seen or were likely to see. William had spent much of the last twenty years in the saddle, his children had learned to ride almost before they could walk and they may all have been forgiven for thinking that they knew all there was to know about horses and horsemanship. But it was the Indians who became their instructors, teaching them how to break in and train a colt and how to perform acrobatic feats in the saddle.

The horse was so essential to the Tehuelches that they had a saying, "A man without a horse is a man without legs", and when asked what they would do if left on the pampa without one they replied, "Sit down." Yet the horse was a comparatively recent addition to their way of life. Animals which escaped from sixteenth-century Spanish settlements on the River Plate proliferated southwards but probably did not reach southern Patagonia, by way of northern tribes, until the late seventeenth century. From a people who stalked their prey on foot until within bow-shot, the Tehuelches became skilled horsemen who hunted on the open pampa. By modern times, their prestige was such that in 1904 the Argentine government sent six Indians from Sierra Palique near the Río Gallegos to the Louisiana Purchase Exposition in St Louis, Missouri, held to celebrate the centenary of the purchase of Louisiana from France: a Tehuelche, Coloko, displayed the boleadora, and two others, Coco and Casenero, were outright winners of the international riding competition.

One of the splendid if increasingly rare sights at Hill Station was of an Indian riding at full gallop. There were few hazards on the open pampa, mostly the holes made by burrowing-owls and by small mole-like rodents, called tuco-tucos because of the repeated grunts they made underground. These could cause a horse to stumble, as

Lady Florence Dixie discovered: "Putting his foot in an unusually deep tuco-tuco hole, my little horse comes with a crash upon his head, and turns completely over on his back, burying me beneath him in a hopeless muddle." On the rare occasion that an Indian fell off his horse, he was expected to land on his feet and was jeered at if he did not.

The Indians had formerly always ridden bareback. They still did so when merely exercising their horses. Each evening they congregated at a track which ran the length of the lake at Kippern Aike. They first cantered side by side and then broke into a full gallop at a given point, there being a wager on the result of the race. Even after a hunt, those owning spare horses would exercise them. On the hunt itself, however, or on the march, they used bolsters or saddle-cloths piled one on top of the other, which also served as blankets at night. A select few possessed an authentic Spanish saddle, bartered — or stolen — from a visitor, but most used a primitive version of their own: two boards, an inch thick, six inches wide and two feet long, were rounded at the corners so as to fit the horse's back and were joined by two strips of board passing across the back, the pieces being lashed together with leather thongs. The bridle and reins were fashioned from guanaco hide, and, when in rocky areas, guanaco-hide shoes were fastened to the hooves. Spurs consisted of sharpened nails driven into a piece of wood which was secured around the ankle with leather thongs, but these were only used on organized hunts; because the horse was so important in life and death, the Indian generally treated the animal more humanely than did the cristianos:

> He'll see it's fed when of food himself
> He's scarce got a single thing.
> To watch it, many a nap he'll miss.
> He's slovenly in everything but this.
> By night he makes his family sleep
> All around it in a ring.

A cristiano's attitude was more likely to be that of the settler Jack Lively:

> It does not do to be too kind to horses out here. They take advantage, then you have to suffer. We have two mottoes in the camp regarding horses and other animals generally: "Better to be sure than sorry" and "Better to count their ribs than their tackle".

In future years the Hallidays were to try to emulate the Indians: when Archie once lent a troop of horses to Lucas Bridges, son of the missionary Thomas Bridges from Tierra del Fuego, for a lengthy ride in the foothills of the Andes, he gave strict instructions that if any of them became too lame to travel, they were to be shot without hesitation, rather than be left with strangers.

When breaking in a colt, the Indians relied on patience and cajolement. Within time, even the children would be able to ride the wildest stallion. If a horse was lame, it was tied up for several hours without food or water, cut deeply in the leg, left tied up again for a few hours and finally turned loose. If its back was saddle-sore, the wound was cleaned with a knife and a type of mineral salt applied, a rare and highly prized substance which was found so high in the cliffs of Santa Cruz that the Indians had to hurl stones to dislodge it.

Most of the horses bought from the Indians by the Hallidays were the progeny of wild mares and criollo studs bartered from a northern tribe. The children named the first horse Indio, and others were to be called Farol, Jerry Blanco, Dick and Christmas. They were the same type of horses on which A.F. Tschiffely was to make his famous 10,000-mile ride the length of the Americas in the mid-1920s, Mancha and Gato formerly belonging to the Tehuelche cacique Liempichúm.

The Hallidays never witnessed the scenes at which Indian horsemanship must have been displayed at its most dramatic: an inter-tribal battle, or a raid or *malón* on a cristiano settlement. By the 1880s the Tehuelches had lost the incentive to challenge former enemies or to resist the cristiano's presence. Although far outnumbering the inhabitants of Punta Arenas in its early days, they had not once attacked the settlement, any trouble there being caused by the cristianos themselves. The northern Tehuelches had shown little aggression to the Welshmen of Chubut, and at Hill Station it was not the Indians who were to show violence but the peons. However, the raids carried out by Patagonian Indians, including, no doubt, some Tehuelches, had formerly been dreaded for their simply and brutal effectiveness, as described by Falkner:

> They would set up camp about 30 or 40 leagues from the enemy and then send out scouts who learn with great exactness every house and farm of the straggling village they intend to attack, also as to give an account of their disposition, the number of their inhabitants and their measure of

defence. The army approaches separately in small bodies. A few hours after midnight, they make the assault, kill all the men who resist and carry away the women and children for slaves. The Indian women follow their husbands, armed with clubs, bowls and sometimes swords; and ravage and plunder the houses of everything they can find that may be of service to them, as clothes, household utensils etc. Thus loaded with booty, they retire as fast as they can, resting neither day nor night, till they are at a great distance.... Here they stop and divide their booty, which is seldom accompanied without great discontent for some or other of them, and these often terminate in quarrels or bloodshed.

After a malón a wasteful slaughter of cattle would often ensue for the sake of the blood. A few of the hides might be used for making toldos, but usually the carcasses were piled up and left to rot. What became known among the cristianos as "Indian-stench" could be smelled for miles and at times became too repugnant even for the Indians themselves, who were forced to move camp.

The Patagonian Indians were renowned for being excellent fighters with a reputation for extreme bravery. A member of Cavendish's expedition recorded: "These Indians are very desperate and careless of their lives to live at their own liberty and freedom." Hudson wrote of two Indians being chased by the Argentine militia: one of them could easily have escaped, "but when his companion was thrown to the ground through his horse falling, the first Indian turned deliberately, sprang to the earth, and standing motionless by the other's side, received the white man's bullets". Darwin wrote of an incident in which four Indians were captured by the Argentine commander General Juan Manuel de Rosas. One of them agreed to reveal the whereabouts of his tribe's encampment on the condition that Rosas first execute his three comrades. This done, he said: "You may now cut my throat also, for I shall never tell you anything." And Musters wrote that when the cacique Cuastro was dying with several bullets in his body, he sprang to his full height and shouted: "I die as I have lived — no cacique orders me." Even Coan admitted: "Divest European cavalry of firearms and they would stand a poor chance before the Patagonian cohorts armed with bolas and knife." Despite their bravery and the advantage of knowing the terrain, the Indians were bound eventually to be overwhelmed by their adversaries' armoury and military expertise.

Although the Hallidays never saw the Indians at war, they were privileged enough to view their remarkable horsemanship on an organized hunt. More usually, the Indians roamed the pampa in pairs or small groups and, with the help of their dogs, chased wild game at random. Sometimes, especially in late summer when the wildlife was in peak condition, a well planned hunt took place, in which the whole tribe was expected to participate. Musters described the preparations:

> The cacique, who has the ordering of the marching and hunting, comes out of his toldo at daylight, sometimes indeed before, and delivers a loud oration, describing the order of the march, the appointed place of hunting, and the general programme; he then exhorts the young men to catch and bring up the horses and be alert in the hunt, enforcing his admonition, by way of a wind up, with a boastful relation of his own deeds of prowess when he was young. Sometimes they breakfast, but the general custom for the men is to wait until the day's hunt has supplied fresh meat. When the cacique's oration — which is very little attended to — is over, the young men and boys lazo and bring up the horses, and the women place on their backs ... the saddles. Others are strapping their belts on, or putting their babies into wicker-work cradles, or rolling up the skins that form the covering of the toldos and placing them on the poles of the baggage-horses; last of all, the small beakers, which are carried on the march, are filled with water ... they then take their baggage-horses in tow and start off in a single file.

When the tribe had arrived at the site of the hunt, two men set out at a gallop around a pre-arranged area of the pampa, lighting intermittent fires to mark their tracks. After a few minutes, two more men set out, and so on, until all the hunters had fanned out into a wide semi-circle which then converged on a line formed by the slowly advancing caravan of women, children and pack-horses. Thrown into panic by the fires, the wildlife found itself trapped. It was not killed within the circle itself, however, where there was total confusion, but was encouraged to try to escape between the hunters who either intercepted it or, if necessary, chased it, making a quick killing and immediately returning to their places in the circle. Dogs were not used as they would only have helped to turn the prey back into the circle, but each Indian took two horses with him, one to ride while forming the semi-circle, the other for the hunt itself, which required a fast and, above all, a fresh mount. When there was apparently no game left inside the circle, two or three of the best hunters were sent inside to scour the area for any rhea that might be hiding there. The dead animals were then collected, some of the

titbits and blood consumed raw on the spot and the rest taken back to the toldos to be divided among the hunters and their families, the children taking first choice. That night, great festivities were likely to take place, to celebrate if the hunt had been a success, to commiserate if it had not.

In hunting, the Tehuelches depended largely on the boleadora, introduced by the Araucanos after the battle of Languiñeo in the eighteenth century. Previously they had relied on the bow and arrow in both the hunting and warfare. When fighting on foot they also used the *chuzo*, a lance up to 18 feet long, ornamented with rhea feathers, with a blade a foot and a half long, capable of penetrating even the coat of mail worn by some of the enemy. They also had a few firearms, obtained from raids on settlements or bartered from neighbouring tribes or from passing traders. Musters wrote of a typical group possessing guns and revolvers "in proportion of about one to every four men". At Kippern Aike, however, there were few rifles or revolvers. The Indians only wanted to hunt, and for hunting the boleadora was still their favourite weapon.

While living in the Falkland Islands, William Halliday had seen the boleadora used by the remaining Argentine gauchos. Kelpers would sometimes practise for fun on the geese, but he himself had never tried his hand. It was of little use on islands devoid of wild quadrupeds, less effective than the lasso for working with horses and cattle, and hardly practicable for sheep. On the wild expanses of Patagonia, however, it was essential for survival.

There were two types of boleadora. The *avestrucera*, used exclusively to capture the rhea, consisted of two balls, or *bolas*, about four inches in diameter and weighing 1 lb., each attached to a leather lanyard or *soga* about five feet long. The *guacheken*, known by the gauchos as the Three Marias (also the name of three stars forming the belt of Orion, one of the most prominent constellations in the sky on the pampas), was used to catch the guanaco and consisted of three slightly heavier bolas joined by two sogas. The best bolas were made of bronze, and the Indians were known to travel many leagues to find suitable ore, but usually they had to be content with stones. At Kippern Aike, it was a question of searching through the scrub or on the shores of the lake for those of an appropriate size, knocking them together until they were perfectly rounded and wrapping them in hide taken from the hock of a guanaco. If a boleadora was to be used for working with cattle and horses, where the intention was to catch and control rather than to maim, the Indians used pieces of old iron cooking-pots beaten together and wrapped in rags, or an excrescence of a birch tree abundant in the Andean foothills, as heavy bolas could easily break an animal's legs.

The Indians taught the Hallidays how to fashion a soga. A guanaco was decapitated, an incision made just above the shoulder and the skin of its neck dragged off in one piece. The wool was picked off, and the skin, softened by hand, was carefully cut into strips and plaited. The sogas used for the smaller avestrucera could be made from the sinews of a rhea: the lower joint of the rhea's leg was dislocated and, with the leg-bone serving as a handle, the sinews were pulled out, dried in the sun, split into equal lengths, laid in a moist place to soften, and then plaited, cooked brains being used to make them more pliable.

To throw a boleadora looked simple. It was not. One of the bolas, known as the *manijera*, was held in the hand and the rest of the device was whirled several times above the head. When enough momentum had been built up for the weapon to cover the required distance, it was thrown at the prey to entangle itself around the animal's legs. All this was usually done at full gallop. The Hallidays were astonished at the ease with which the Indians could "ball" a rhea or guanaco at distances of up to 70 yards, sometimes even throwing the weapon from under the horse's neck. A well trained mount could carry its master out of danger even if its fore or hind legs had been accidentally trussed together by the bolas misfiring.

It was easy to see why an adversary had to be alert in time of war:

> When you're drawing off, and you think he's through
> He'll send a chance shot after you
> And if it catches you, for sure,
> It lays you out for ever.

When the Hallidays first tried the weapon, they found it as difficult as "casting a fly at full gallop". At first they practised on foot, using a chosen shrub as target. They then balled the same shrub from horseback. Finally they tried balling a rhea, learning the nasty consequence of wrapping the bolas round the legs of one's own horse. They also discovered that, if not balled properly, an animal simply ran off with the boleadora and, unless the hunter had an excellent horse, the weapon was lost for ever. There was a need for the closest co-operation between horse and rider, and it was almost impossible to use the weapon when mounted on a horse unaccustomed to the whirling of the bolas above its head. But, at last, some prey was brought down and killed. It was an important moment for the Hallidays. They would never be hungry again.

ONCE THEY HAD mastered the boleadora, the Hallidays could at last hunt the two most prolific wild creatures of Patagonia, the rheas and the guanacos. The former, referred to by most settlers and visitors as ''ostriches'', had been known and described from the sixteenth century onwards but were only given the name rheas in 1752 by the ornithologist P.H.G. Möhring; no one was sure why he called them after the Greek goddess, wife of Saturn, unless perhaps because of their inherent grace. Previously they had been known as the *ñandú*, from the Brazilian, or simply as American ostriches. They in fact differed in several ways from their African cousins, being on average about five rather than eight feet tall, having three instead of two toes on each foot and with a wing-plumage which, though not quite as luxuriant, was more developed, covering the whole body when the wings were folded. Flocks of rheas had formerly roamed much of the Argentine pampa as far north as the Bolivian and Paraguayan borders. It was Darwin who noticed that there was a striking difference between the bird of the north and that of Patagonia. The former, *rhea americana*, would run with its wings outstretched like sails and its head held high, the latter with its wings closed and the neck held almost horizontally, making the bird appear even smaller than it was. He sent a specimen from Port Desire to the Zoological Society in London, who in 1837 appropriately classified the Patagonian bird as *rhea darwinii*.

The rhea had fewer peculiarities, both real and fabled, than the ostrich. Able to distinguish between seeing and being seen, it would not bury its head in the sand. But, in its own way, it was a strange enough creature to be an apt inhabitant of Patagonia. Its very presence, like that of humming-birds and parakeets in the Straits, could surprise a traveller who associated the ostrich family with hotter climates. The warning cry of the male, ''a repetitive sharp hissing note, pí, pí, pí", was a baffling sound, as Darwin recalled: "I thought it was made by some wild beast, for it is a sound that one cannot tell whence it comes, nor from how far distant." The nest, merely a shallow hole about two and half feet in diameter scooped out of the earth, was a welcome discovery at Hill Station because, in early summer, it could contain up to forty large eggs. It was the male bird, larger, darker and faster, which sat on the nest for the incubation of three weeks, the grass all around being eaten bare. Though willing to accommodate the eggs of several females, it was said somehow to make sure that those of its own mate were well-protected and were not among those, known as *huachos*, which became separated from the nest and were never hatched. If molested, the bird would draw the predator away from the nest by limping as if wounded, and if it returned to find a single egg missing it was known to smash all the others to smithereens in its fury and grief. The rhea had several enemies: the puma ate the bird and the eggs, as did the wild cat, more common in the Andes and rarely seen at Hill Station; the fox devoured the chicks and the eggs, while the condor and carancho ate only the chicks. The most dangerous predator of all, however, was man.

The rhea or *oyne* had always been the Tehuelches' favourite food. Drake's expedition found at San Julián "great store of ostriches at least to the number of 50, with much other foule; some dried and some in drying for the provision, as it seemed, to carry with them to the place of their dwellings". If the bird was in prime condition, virtually all of its flesh was eaten, from the breast to the gizzard, large enough to fill both hands. The Indians even sucked out the eyes. Cooked in Tehuelche style, the "wee ostrich" became a popular dish at Hill Station. The recipe was as follows:

> Take a good fat picana, being careful that the skin has not been torn by the dogs. Take the bone out and cut the meat into thin steaks leaving one good thick steak on the skin. Put a dozen stones — about the size of a child's fist — into the fire. When thoroughly hot, spread the skin flat on the ground and lay the stones on it. Cover the stones with steaks and make a few slits in the skin on the outside. Thread a strip of skin through the slits and draw tight in the shape of a bag. In about an hour's time, there are twelve pounds of meat perfectly cooked, and not a drop of gravy has been lost. It's grand eating, fit for an alderman. (Lively)

This "stoned ostrich" or *asado con piedras* met with unqualified approval from settlers and explorers. Lady Florence Dixie, "though

somewhat of a gourmande", declared it to be the best meat in the world, and Beerbohm, more used to the cuisine of the Café Royal in London, conceded that ostrich wings were "the greatest delicacy, tasting somewhat like turkey and, as I then thought at least, perhaps even finer". Rhea eggs, too, were a popular part of the Patagonian diet. Preferably eaten fresh in late September or October, they remained edible for a long time afterwards. The Hallidays learned, like Hudson, to tell the age of an egg by its colour:

> When newly laid, the eggs are of a deep rich green, and the shell possesses a fine polish. They very soon fade, however; at first the side exposed to the sun assumes a dull pale mottled green; this colour again changes to a yellowish-green and again to a pale stone-blue, becoming at last white.

The method of cooking the egg, simple in theory, could be precarious in practice:

> You break a small round hole in the top of the egg and, after having removed some of the white, which is rather heavy on the stomach, and having thoroughly beaten up the yolk, you set the egg on its end in the ashes, at a little distance from the fire, carefully turning it over now and then to prevent the shell from cracking. Whilst cooking, it must be occasionally removed from the fire, and the batter stirred well, or else it will stick to the sides of the shell and burn. In a quarter of an hour, it will be well roasted; add pepper and salt, if you have any, and serve.

Placing the egg near the fire, turning it over and eventually removing it from the ashes required nimbleness and could result in many a broken egg and burnt finger. It was also possible to boil the egg, but, as it was about ten times the size of a chicken's egg, the procedure took some forty minutes and required a truly Patagonian egg-timer. The Hallidays had also been warned, and soon learned, that the eggs could be indigestible. Two eaten in a day were thought to endanger a man's life, but Beerbohm managed three or four, and his companion, Guillaume, once safely ate six in the spate of eight hours, together with his ordinary meals.

If a rhea was thin and not fit for eating, its skin was used by the Indians as a blanket. This was often infested with vermin and Musters gave a warning to anyone visiting the toldos: "Never allow the squaw of the establishment to place ostrich-mantles under your sleeping hides." The sinews from the legs were fashioned into sogas or used as thread. The skin of the neck, pulled off like a glove and worked with oatmeal, could be made into a bag or, as Grandpa

51

52

McCall discovered to his pleasure, into an excellent tobacco-pouch. Lady Florence Dixie was once about to throw away the carcass of a rhea when she remembered that its leg-bones made "very nice handles for umbrellas or whips". And, again exemplifying the importance in Patagonia of using to the full whatever Nature offered, the Indians even manufactured buttons from the quills of the feathers: the quill was split lengthwise and the strips plaited, the ends being tucked in so as to be invisible.

As far as the Tehuelches were concerned, however, the greatest value of the oyne lay not in its flesh or bones but in its plumage, which was both popular for personal ornamentation and essential for trade. As commerce between Indian and cristiano developed, demand for "ostrich-feathers" soared, and a massive slaughter of the bird took place. The consequences were particularly disastrous as it was the feathers of the young bird or *chara* which fetched the highest prices. In the middle of the nineteenth century, it had still been possible to claim that, were it not for the Indian hunters, the rhea might have overrun large areas of Patagonia. By 1871, however, Hudson was lamenting that, despite engaging a dozen Indians to hunt for him along the Río Negro, "they failed to capture a single adult bird". By the turn of the century, the ornithologist Alfred Newton wrote: "For the purpose of trade in them, it is annually killed by the thousands so that it has been extricated from much of the country it formerly inhabited and its total extinction as wild animal is probably only a question of time." The Indians cannot be held totally responsible for this drastic decline. Cristiano hunters or *chulenguadores* were soon at work. By the 1890s there were at least eight full-time professional ostrich-hunters in the Río Gallegos area alone. Added to this, the rhea was destroyed by the sheepfarmers. Each October, as in Egging Week in the Falkland Islands, the Hallidays went in search of ostrich-eggs and a prize was offered to the member of the family who found the nest containing the largest number. The children also kept a secret eye open for "the fabulous pure-white ostrich" which, the Indians assured them, was to be seen on the rarest occasions roaming the camps of Hill Station. It was on no account to be killed, nor its eggs touched.

Killing a rhea was not always easy because, in the face of so many natural enemies, it had built up strong defences. Its chief asset was speed. A chara could run at a reasonable pace almost immediately after it had been hatched and a fully grown bird reached speeds of up to 40 m.p.h. Francis Fletcher remarked: "They runne so swiftly and take so long strides that it is not possible for a man in running by any means to take them, neither yet to come so nigh them as to have any shot at them with bow or peece." When disturbed, a rhea would run straight as a die at top speed for several hundred yards, then, as it began to tire, it constantly changed directions, darting hither and thither until, quite suddenly, it would squat down and be uncannily camouflaged amoung the greys and browns of the pampa. Its powerful eyesight and range of vision protected it from stalkers, and just as the African ostrich sought the protection of herds of zebra and antelope, so the rhea had learned to mingle with the guanacos, and soon with the flocks of sheep.

Before the days of horse and boleadora, the Indians had used intricate tricks to catch the speedy oyne. Preferring to hunt them in winter when their eyes were dazzled by the glare of the snow and their speed was reduced by saturated plumage, they would drive them into a river where their legs froze so that when they drifted on to the shore they were easily captured, since they were unable to move. Human decoys were sometimes used, as witnessed by Drake's expedition:

> They have a great and large plume of feathers orderly compact together upon the end of a staff; in the forepart bearing the likeness of the head, neck and bulk of an ostrich; and in the hinder spreading itself out very large, sufficient — being holden before him — to hide the most part of the body of a man; with this they staulke, driving the birds into some straits or neck of land close to the seaside, where spreading long and strong nets with their dogs... they overthrow and make a common quarry.

With the use of horse and boleadora, however, no such strategems were necessary. Although running faster than a horse at full gallop, the rhea had less stamina, and as long as the hunter did not lose sight of it he would eventually be able to entangle the bolas round the long, spindly legs. He could then shoot it through the head or, like the Tehuelches, seize it by the throat and swing it violently around until the neck broke. The hunt was often a long one, and the Hallidays were agreed that there was nothing more exciting about life in Patagonia than "giving chase to the wee ostrich":

> 21 May 1901: Camp Río Bote: followed the hills down until struck Río Santa Cruz.... I saw ostriches — large ones. I let the dogs go. Lynch and Corbata took after one, and Chimango after the other, Lynch and

Corbata soon caught theirs, Chimango failed. I came across a flock of charas.... We all chased. Lynch caught one by himself, and the other dogs managed to snaffle another.

The guanacos, or "camel-deer" as the Halliday children called them, were even more abundant at Hill Station than the rheas and, as voracious herbivores, posed a great threat. Strange creatures, they at first mystified the Hallidays as much as they had the early navigators, some of whom referred to them as "stagges and bufflas", others as "harts with very long necks", "winnackews", "sheep as big as small mules" or "Spanish sheep". An apparent amalgam of different animals, they had, as Musters put it, "the neigh of a horse, the wool of a sheep, the neck of a camel, the feet of a deer, and the swiftness of the devil". The males' warning cry was a mixture of "a bleat, a laugh and a neigh". On one hand they were very much members of the camel family, like their northern cousins the llama, alpaca and vicuña, with the same long neck, thick wool, indolent expression and undulating throat. On the other hand, they looked like deer, standing four feet high at the shoulder, four to five feet long from the tip of the nose to the tail, with a straight back, long legs and considerable grace which impressed, among others, the Chilean poet Pablo Neruda in his poem "Algunas Bestias":

> ... the guanaco, light as oxygen,
> in his golden shoes
> trod the wide brown uplands.

The ambiguity of the guanacos' appearance was matched by some of their habits. As the Hallidays had already discovered to their advantage, the animals deposited their droppings in communal heaps, a practice not only useful to man, who burned the dung as fuel, but also to the rhea which liked to feed off the insect-ridden heaps and to the guanacos themselves, as the grass in the area became thick and lush. The settlers were to discover an even stranger habit.

In a cañadón on the "road" to Cape Fairweather, they came across an enormous pile of bones: "It was as if someone had gone round the camp and tidied up the skeletons." The children conjectured in their own way: the camel-deer, like the elephants, preferred to die among those of their own kind, or the bones had been dragged there by the cursèd lions, or perhaps they were not animals' bones at all and this was a graveyard of the giants! The true explanation was that the guanacos instinctively sought a sheltered place in which to die.

When ascending the Río Santa Cruz, Darwin noticed that a wounded guanaco invariably headed for the banks of the river and preferably to a place where there was plenty of vegetation. At one such spot he counted twenty skulls and observed that the bones had not been gnawed or broken as if dragged there by pumas: "The animals in most cases must have crawled before dying beneath and amongst the bushes." On occasions, especially after a bad winter, the piles of bones found at Hill Station were huge. A blizzard would force whole herds to seek shelter, and there the snow would pile up around them as they huddled together, too cold to attempt escape. They soon ate the bushes bare and, as Lucas Bridges described it, "with a white world of snow stretching for leagues in every direction, they give up hope and one by one lie down to die." William must have thought ruefully that the same fate might all too easily befall a flock of sheep.

Despite such catastrophes, the guanacos were still widespread in many parts of Patagonia. They usually roamed in small groups consisting of one adult male and between four to ten females, or exclusively of young males which lived separately until maturity. In severe weather, however, they congregated into much larger herds, and it was then that it became clear just how numerous they were. An early English navigator wrote of "drives of six and seven hundred", and Darwin of a herd of one thousand. During a month's severe flooding of the Río Gallegos in 1899, John Scott recorded: "More amazing still were the vast numbers of guanaco which the flood carried before it. They were so numerous that those who had not seen them with their own eyes could scarcely credit the tale that they were as dense as trees." Jack Lively, trapped on the pampa in the winter of 1901, wrote: "I saw a guanaco put his head over a ridge, then another and another. There were literally thousands of them."

These large numbers existed despite the fact that it was a principal prey of both the puma and the Indian. Although, unlike the llama, it was never used as beast of burden, the *nau* had for centuries been as vital to the Tehuelches as the bison was to the Indians of North America. It was hunted both for its flesh and for its hide. The meat, blood and offal of the young *chulengo* were considered delicacies, as were its marrow-bones and the curdled rennet found in its intestines. The adult was killed for food only if the Indians intended to stay in one place for a long time or simply felt inclined for blood. Some cristianos praised the taste of even the fully-grown male, Musters comparing it to "very tender veal", Darwin writing that an enormous

53

54

117 lb. animal he ate for Christmas dinner 1833 was "excellent". A nourishing soup could be made out of a guanaco's head, with the addition of rice, dried vegetables and chili, and Lady Florence Dixie paid special tribute to the tongue, brain and "fat-behind-the-eye, the ne plus ultra of pampa delicacies". When Mary eventually managed to procure some salt, she learned from the Tehuelche squaws how to make guanaco pemmican, useful for a rainy day: "The haunches are charqueared — the meat cut off in thin slices and, after a little salt has been sprinkled over it, dried in the sun. When dried, it is roasted in ashes, pounded between stones and mixed with ostrich or other grease." A mere handful could satisfy the appetite for days.

But it was principally for their hides that the Indians slaughtered the guanacos. If in good condition, they were used for making mantles, ponchos, blankets and canvases for the toldos. If in poor condition, they were used as saddle-cloths. The skin of the neck, particularly durable, was fashioned into boots, lassos, bridles, sogas and the covering for bolas. The skins from the legs were used as saddle-bags. Like those of the rhea, the tendons served as thread, and the thighbone was carved into a musical instrument or cut to make dice.

To skin a guanaco by the Tehuelche method was a laborious task, the thumb or a flint knife being used to prise the hide from the carcass. Once removed, the skin had to be exposed to the air for a few minutes before being folded or it would tear easily, and only the fur side could be dried in front of the fire as the skin itself was likely to crack. The young chulengo provided the softest and most treasured hides, and, each November, a special hunt took place. The skins were left to dry overnight under the usually clear spring skies and then painstakingly sewn together by the women with the use of a wooden puncher, the animals' own tendons serving as thread. From fifteen to fifty chulengo pelts were used for each mantle, which then served as clothing during the day and as a blanket at night.

If the guanaco had been hunted solely for domestic requirements its numbers would not have been so drastically reduced. But the chulengo blanket or *quillango* became almost as important as ostrich-feathers in the Indians' bartering with cristiano traders (as did blankets made from foxes, when between thirty and forty skins were used, and from skunks, where between fifty or sixty skins were used). The highest prices were paid for skins from animals under two months old and, above all, for those of the unborn foetus or *umbro*, whose wool was almost as soft as the famed vicuña's. The pregnant females were killed in their thousands, for they could not maintain their speed for long because of their weight and were easily killed: "One man not infrequently captured five or six in the chase, extracting the young and taking the skin for mantles and the carcass for food, while the hide of the mother served, if necessary, to repair the toldos." Professional cristiano hunters joined in the hunt. By the 1890s, an estimated 2,000 quillangos were being traded annually in Punta Arenas, thus entailing the slaughter of 50-70,000 animals.

Sheepfarmers also killed them because a guanaco ate twice as much grass as did a sheep. Some employed Indians to "clean" their camps of the animal. Others, like the Hallidays, did the hunting for themselves. Sometimes the killing was accidental. When fences were erected on the farms, the guanacos were at first unable to jump them and were frequently trapped by the weather, though by the twentieth century they had so successfully acquired the art of jumping that ten-foot-high fences had to be erected around any special paddock.

To kill a guanaco was not always easy for, like the rhea, it could run with remarkable speed. When only three days old, the chulengo could match the pace of a horse's hard gallop and, within a month, was almost unapproachable, a few leaps taking it away far beyond the speed of a horse. As it ran, it would constantly duck and sway its head. It was also difficult to stalk as it had strong powers of eyesight and smell, and, except to the experienced eye, its spoor resembled that left by a caravan of Indians. It could also be vicious when cornered, lunging out with its powerful hooves, biting hard and, like other members of the camel family, spitting its foul-smelling saliva to a distance of two or three feet. The Hallidays sometimes came across a place where a guanaco and a puma had fought to the death and, though the puma invariably won, the struggles had evidently been bitter ones. Examining a guanaco she had killed, Lady Florence Dixie found that a bullet had passed through its lungs and lights and had lodged near the spine, "and yet thus severely wounded, he had gone quite ten miles at a cracking pace". The Hallidays once saw a guanaco with a broken leg plunge into the estuary and swim across to the other side.

Despite its initial timidity and ultimate bravery, the camel-deer had a characteristic which often proved its downfall, that of inquisitiveness. In the days when the Indians relied on bow and arrow and found it difficult to approach close enough for the flint

arrow-heads to pierce the thick wool, they played on the animals'
prying nature. Pigafetta wrote: "When these people wish to catch
some of those animals, they tie one of the young ones to a thornbush.
Thereupon, the large ones come to play with the little ones; and these
people kill them with their arrows from their place of concealment."
Darwin noticed that, "If a person lies on the ground and plays
strange antics, such as throwing up his feet in the air, they will almost
always approach to reconnoitre him." Rather like the Indians
themselves, the guanacos at first kept their distance but eventually
could not resist the attraction of a new settlement. Richard
Coppinger, an Irish naval surgeon and naturalist who took part in an
exploring cruise in Patagonian waters in 1879, noticed a herd of
guanacos on the very tip of Dungeness Point, "browsing quietly near
the beach, as if a passing ship were an object familiar to their eyes".
Some hunters claimed that, although terrified by gunfire, the
guanaco would eventually investigate the area from which the
gunfire had come. The hunter had only to wait.

Although the guanaco's speed and courage gave it some chance
against the puma, it could not hope to resist the combined force of
rifle and boleadora as used by Indian, chulenguador and
sheepfarmer. The Hallidays first helped rid the guanaco of its
dreaded enemy, the puma. They then assisted the virtual
disappearance from Hill Station of the camel-deer itself.

IN FEBRUARY 1886 two important events took place at Hill Station: the first wool-ship called in from Punta Arenas, and the Hallidays' cousins, the Rudds, arrived from the Falklands.

It was at the sight of a ship steaming into the estuary that the settlers realized how isolated they had been during their first seven months in Patagonia. The only vessel to call during 1885 had been the Argentine cruiser *Villarino*, under Captain Federico Spurr, which brought materials for the construction of a sub-prefecture on the south bank, officially inaugurated on 19 December. Representing the only official communication between Buenos Aires and southern Patagonia, she was in poor condition and of small tonnage, taking up to sixty days to make the round trip and symbolizing the indifference that the federal capital showed towards its southern territories. It was not until 1898 that the authorities resolved to establish a regular steamer-service, and not till 1910 that the service came into effect.

The arrival of any vessel caused great commotion at Hill Station. At the cry of "A ship!" all work stopped and the children ran to the top of Square Hill or down on to the beach to see where she would drop anchor and which flag she was flying. Until the sub-prefecture was completed and a town began to emerge, the visit of a ship gave the settlers not only a chance to replenish provisions but also rare contact with the outside world. In Punta Arenas, which by Patagonian standards and certainly when compared to Río Gallegos was a busy port, the colonists would abandon even their Sunday horse-races at the sight of a ship entering the harbour, all anxious for anything new or strange. The first wool-ship to visit Hill Station was a small schooner belonging to Bráun y Blanchard. She lay on the beach at low tide and the bales were loaded directly by hand, taken to Punta Arenas and from there to Buenos Aires to be auctioned. In the years ahead, the visit of a wool-ship would take on ever more importance for, like most other settlers, the Hallidays eventually decided to sell their wool directly to Britain.

In February 1886 the *Villarino* returned with more building materials and also a captain for the port, *el capitán* Manzano, or "Captain Apple" as the Halliday children called him. The following month a small steamer arrived from Punta Arenas carrying on board the Hallidays' cousins, the Rudds. Like William, Jack Rudd had been working on the Falkland Islands as a shepherd under contract with FIC and had become increasingly concerned about his family's future on the islands. He also had a personal incentive for leaving because his brother William, another contract shepherd, had been murdered at Blakeley's near San Carlos by an Argentine gaucho who eventually hanged himself from a cliff. When William Halliday returned to the islands after his exploratory visit to Patagonia, Jack expressed an interest in taking out a similar lease of land and the most attractive and logical choice was the camp adjoining Hill Station at Cape Fairweather. Towards the end of January 1886 he settled his affairs and set out from his home at Dos Lomas, eight miles west of Darwin, with his wife Anne McCall, who was Mary Halliday's sister, and their four young children, Agnes, aged seven, Helen, five, Mary Jane, three, and the one-year-old John Rodgers. They rode to Port Stanley, embarked for Punta Arenas aboard SS *Minas* of the Kosmos line, and from there to Río Gallegos. On 22 February they crossed over to the north bank without incident. Patagonia gave them a warmer welcome than it had done to their cousins, who watched anxiously and excitedly at the prospect of provisions, company and gossip.

There were now more mouths to feed, but more hands to help work the farm. Both families knew that it was in their interest to establish Hill Station efficiently before the Rudds moved "down to the Cape". Besides, Jack still had to ride north to Puerto Santa Cruz to formalize the lease of his land. As the Hallidays had up to this time been living in their makeshift accommodation, the first priority was to build a house, and this presented special problems: there was no natural timber, no bricks, no tiles and no glass, yet the building had to be solid enough to withstand strong winds, torrential rains and heavy snows. The Rudds had brought some planks and corrugated-iron sheets from the Falklands and, supplementing these with driftwood from the beach and with

material taken from a shipwreck on a small island just off the Cape, the settlers built a simple four-roomed, single-storey house, 23 feet by 24, with shingled boards on the outside and grooved boards on the inside, stuffed with grass between them for insulation. The corrugated iron was used for the roofing. The house was positioned some 400 yards up from the beach above the cañadón in which the Hallidays had spent their first months. At the back, they built a carpenter's shed, a meat-house and a privy. William had also tried to remove a kitchen-stove from the shipwreck, "but when he and Willie went to lower it, it was too heavy for them, and fell deep into the water for ever". Everyone in the family played his or her part in the building. Mendy, for example, was expected to hammer out any bent nails. Jack Rudd was the principal carpenter, as he "had had enough of sheep" in the Falkland Islands, and it was his wife Anne who now, and later on down at the Cape, preferred to work as shepherd and shearer. As soon as they had enough money and materials, the Hallidays planned to add an extra storey and a gallery facing the estuary. But, for the time being, this was home, and to them it was nothing less than a palace. On the opposite side of the cañadón and closer to the beach, they built the first shearing-shed. This meant that the second year's shearing, though still done by hand, would take place under a roof and with a proper fleecing-table. More importantly, the bales would not be left to lie on the beach at the mercy of the weather, and perhaps of thieving Indians, but could be safely stored in the shed until the arrival of the next wool-ship. A sheep-dip was made with timber from a wrecked schooner found north of the Cape, whose ribs also provided posts with which to make sturdier corrals and sheep-pens.

In 1888, the settlers sold their wool for the first time directly to British wool-merchants. Ships of the Glasgow companies of Thom & Cameron, and Spearing & Waldron, began to make annual visits to Patagonian ports, and the captains of such vessels as the *Crossowen*, the *Ruth Waldron* and the *Bootle* became popular guests, for their visits brought not only cash and provisions but news from home. They were emissaries of a world which the Halliday and Rudd parents had left voluntarily and which their children had never seen, but about which they thought and conjectured, especially when times were hard. Although a letter-carrier began to travel the length of Santa Cruz in 1892, the service was unreliable and the settlers would for many years rely on the wool-ships for even the most startling news: "Feb. 24 1901: Capt. Oliver left; washed all day. Boys carted wool. Heard that the old Queen is dead." Such news took even longer to reach farms further from the ports: Evan Noble, working at Cañadón de las Vacas, halfway between the Rios Coyle and Santa Cruz, waited six months before hearing of Queen Victoria's death.

The first British ship to enter the estuary in 1888 was the *Crossowen*. Her skipper, Captain Brazier, was to become a frequent visitor. The Hallidays purchased a small rowing-boat from him. A few years later his wife gave birth to a son while the *Crossowen* was in Puerto Santa Cruz, and he himself so liked Patagonia that, when forced to retire early after losing an eye in an accident, he settled at Pescadores, near Santa Cruz. An oaktree which still stands at Kilik Aike Norte grew from a seedling brought on the *Crossowen*. Captain Augustus Jones of the Spearing & Waldron vessel *Martha Gale* was particularly popular as he and his wife always brought presents from Britain, Mrs Jones once giving two black hens and a rooster for Mary and a musical box for the children. In 1892 they also brought an invaluable spinning-wheel. Mabel Halliday, born at Hill Station on 23 March 1888, was later to remember the Joneses for a different reason. The year was 1894, and the captain and his wife were guests at the farm. Coming across an old pipe that Grandpa McCall had left lying around, Mabel filched some tobacco and crept away to the calafate bushes: "I had a great smoke to myself, and an enormous feed of calafates." On her return to the house, her mother at once remarked on how pale she was looking and made her lie down: "Mrs Jones then gave me my first-ever dose of Eno's Fruit Salts!"

Vessels other than wool-ships rarely called in at Río Gallegos during those early years, but in 1888 the Scotsmen received a special visitor, HMS *Basilisk*, on an official tour of the Patagonian ports to give advice to British settlers on what to do should war break out between Argentina and Chile as a result of the frontier-disputes. The officers asked William to lend them horses and to act as guide on an ostrich-hunt. As was usual on such occasions, the sailors had done little riding lately and they suffered a great deal, especially on their way back when they lost the water-barrel. The children were delighted at being given a length of black ribbon embroidered with "HMS Basilisk". And William and Jack were reassured to know that, in this far away place, they and their families were not completely neglected by Her Majesty's Royal Navy.

The visit of a ship to the Río Gallegos was sometimes unscheduled and a disaster. Many sailors experienced the treachery of the estuary as the Hallidays had done on the day of their arrival. The shipwreck off the Cape was proof enough. And on the small island was a grave surrounded by a chain, with the inscription:

> Walter Deputten
> Native of Hanover.
> Drowned coming from the mainland to the island.

The German had been sheltering on the island with other shipwrecked sailors, believing the Indians to be scared of water. William himself was to learn the dangers of those hundred yards to the island. Although an excellent swimmer, he once nearly drowned while visiting the shipwreck: "His dog called Sluster kept trying to scramble on his back." Even Captain Brazier was betrayed by an estuary which he knew as intimately as any man alive: the *Crossowen* was once inadvertently grounded on the south bank opposite Kilik Aike Norte and her captain had to wait with some embarrassment for the next spring tide until he could dislodge her, the place still being known as Brazier's Point. A Spearing & Waldron wool-ship, the brigantine *Bootle*, was to suffer more calamitously. Late in 1897 she was on her way south to Punta Arenas with a cargo of barrel-staves. On Boxing Day a sudden gale forced her captain to drop anchor in the north channel off the Cape, but she was given too much cable and stove in. The crew of fifteen, mostly Norwegians but under a British captain, scrambled ashore to the Rudds' farm. They had evidently been celebrating Christmas too heartily, for they had mistaken Cape Fairweather for Dungeness Point at the eastern entrance to the Straits. When they arrived at the farmhouse carrying their logbook and a huge Union Jack, they argued heatedly that this was indeed Dungeness and, being still far from sober, suggested to Jack that he should leave navigation to sailors and go tend his sheep. In the sobriety of the following morning, however, the captain realized his mistake and, climbing Sugar Loaf Hill and lying stretched out on the Union Jack with the Rudd family enjoying a picnic, he was heard to mutter: "I wish to God she would sink out of sight." His prayers were answered, but not till the 1940s, before which time the hulk of the *Bootle* was still to be seen at exceptionally low tides.

There were to be other incidents. In 1898 an old wooden sailing-ship, the *Annita*, was on her way from Liverpool to Punta Arenas

Margaret Glen

carrying a general cargo. She began to leak so badly in the South Atlantic that her crew had to pump her continuously for the rest of the journey. One of the sailors, a Swede called Gus who later worked for some years as a shepherd at the Cape, told how the *Annita* became so full of water that, by the time she reached the Río Gallegos, her captain wanted to beach her at Cañadón Palo, just north of the Cape. But the first mate, already aware of the dangers of the coastline, persuaded her captain to continue into the estuary, and they successfully reached port. But Patagonia had the last laugh. While the *Annita* was being unloaded, a block of wood fell on the first mate's head and killed him outright. The *Annita* herself was abandoned as a worthless hulk.

In September 1900 the Norwegian vessel *Ostawa*, under Captain Olsen, ran aground on sandbanks opposite the Cape. She finally managed to limp into Río Gallegos, but not before the Rudd children had smuggled themselves aboard and requisitioned a Norwegian flag, which they proudly hoisted from the top of Sugar Loaf Hill as she sailed by. In 1912 the *Margery Glen* was stranded on the southern beach for eight days with a cargo of 3,000 tons of coal. When asked for assistance, the port authorities opened the hatches so that air infiltrated and the ship took fire. She floated up and down the estuary for several days burning all the while, until she was beached on Punta Loyola. Survivors of shipwrecks along the Atlantic coast sometimes arrived at Río Gallegos as the nearest point of civilization. Some came by foot, such as the captain and seven sailors of the sailing-ship *Emma* from the Falkland Islands, which sank some 15 miles north of the mouth of the Río Coyle in September 1895. Others arrived in the ships of their rescuers: "September 1896, disembarked off the *Villarino* 29 men, including the captain of the English frigate *Columba*, which sank September 12 between Cabo Blanco and Tres Puntas."

The visit of a ship, scheduled or unscheduled, was still rare during the early years of Hill Station. By 1896, when the town of Río Gallegos had become capital of the territory of Santa Cruz, the total number of national and foreign vessels calling in at the port was fifty-two, with a displacement of under 25,000 tons. There was one ramshackle warehouse some 500 yards from the beach. There were no skilled dockers, no ship-repair service and no medical facilities. The new captain of the port, Federico Corvetto, complained bitterly but vainly to the authorities in Buenos Aires. When obliged to impose a fine on the captains of the *Bootle* and *Crossowen* for failing to register with the consulate in London, he was given the reply that the consulate had no idea of the existence of a port at Río Gallegos, let alone of any port authorities.

In February 1888, by which time Jack Rudd had finalized the lease of eleven square leagues at the Cape and the flocks had increased sufficiently for William to sell him a hundred ewes and several rams, the Rudds built a proper track and moved down to the Cape to establish their own farm. But they soon ran into trouble. A Spaniard, Victoriano Rivera, rode in from the north to inform them that the shack they had built in which to live until the main house was completed lay on his property. Rivera was a member of a consortium which had purchased a vast tract between the Rios Coyle and Gallegos, comprising the estancias la Delfina, La Angelina and Bahía, the last of which adjoined the Halliday and Rudd properties to the north. The authorities in Puerto Santa Cruz had neglected to tell Jack that Rivera had reserved a strip of land which ran down from Bahía to the estuary, giving him vital access to the wool-ships. The Rudds had no option but to move and rebuild their shack closer to the Atlantic coast.

This type of misunderstanding was common in Patagonia. As the Argentine historian Edelmiro Falcón put it: "Though barren by nature, Patagonia is most fertile in quarrels." As most of the land had never been properly surveyed, demarcation was largely based on lines of longitude and latitude, and exact boundaries were difficult to prove. Matters were aggravated firstly when all the original deeds issued to settlers in the territory of Santa Cruz were destroyed in a fire at Los Misioneros late in 1888, secondly when it became government policy to grant a maximum of eight square leagues to each settler, with the result that individual lessees used pseudonyms or names of relatives in order to lease more land than was allowed. Some farmers took the law into their own hands, more in the style of the North American Wild West. Settlers who wanted to take proper legal action had to appeal to the Governor directly or, by the late 1890s, to the court-house or *juzgado* in Río Gallegos. Any appeal against a local decision was made at *la Cámara Federal* in La Plata, over a thousand miles to the north.

Some disputes between neighbours lingered for years. Every farmer needed an area of low-lying ground on which to shelter his flocks and, if none existed on his own property, it was tempting to

59

60

61

trespass on to someone else's. Arguments ensued over legal boundaries and the ownership of sheep, with accusations and counter-accusations of the deliberate mixing of flocks. Apart from initial misunderstandings with Rivera, the Hallidays and Rudds were fortunate in having no problems with their neighbours. Other settlers were more unlucky. A typical dispute was that between the Frenchman Auguste Guillaume at Coy Aike and the Scotsman Henry Jamieson at Moy Aike on the Río Coyle. First mention of a confrontation appears in Jamieson's diary in May 1893: "Revised the 305 sheep that G—— had delivered. Claimed sheep belonging to me as his property. He painted them red around the rumps." A few days later: "G—— and the chief of police came," but evidently to no avail for on 9 January the following year, Jamieson was still complaining: "G's scabby wethers came up to the point of Moy Aike and mixed with mine." The police were called in again and ordered both farmers to gather in all their sheep for examination. In October 1894, Jamieson was still writing: "A flock of about 2,000 of G's sheep have been three days on my camp — also about 60 mares in front of my house," and in January 1895: "G sent me word that there were about 400 of my sheep close to the corral made for marking lambs. Went to bring them with the dogs but found nothing except a ewe which I was taking off an island when G arrived and kicked up a row but would not listen to my explanation. He appeared to me to be more like a lunatic than a sane man." In June 1896, however, Jamieson was inside a cell of the Río Gallegos police-station. Fourteen of his friends, mostly compatriots and including William McCall, William Halliday and Jack Rudd, immediately made a declaration to the authorities about what they considered an unjust arrest. They threatened to take the law into their own hands. Jamieson was set free. The dispute re-emerged briefly the following year: "May 1897: A lot of G's sheep mixed with mine again on pampa, some of them very scabby." But the Frenchman had had enough. Late in 1897, he packed his bags and moved his flocks to La Julia on the Río Santa Cruz near Lago Argentino.

Some of the disputes were to end in strange ways. According to a story told by E.M. Davies, formerly manager of the estancia Condor and at the time manager of Kilik Aike Sud:

In the summer of 1910 Davies was gathering in a flock of some two thousand ewes when he noticed that several of the animals' ears had been tampered with. The farm ear-mark of a single notch had been cut away and changed to a fore-quarter. He parted the animals, counted 350 with the new mark and at once informed the Río Gallegos police. The paddock in which the ewes had been grazing ran alongside the road to the Gallegos freezer (established at the turn of the century) and the police assumed that the culprit would at a later stage claim that his sheep had strayed. But no flocks with such a mark were expected at the freezer for some time and all the farmers concerned were beyond suspicion, so the theory was dismissed. The untampered sheep were left in the pens overnight and, on the following morning, over a hundred of them were discovered to have the new mark, all in the right ear and all cut clearly as if with an implement. There was no trace of blood on any of the animals which suggested that they had not been bunched up into a race or small pen, but had been dealt with individually in the main pen. But who in his right mind would go round a large pen catching sheep in the dark when a smaller pen was available? And how did the criminal escape detection in a comparatively compact farm such as Kilik Aike Sud? The dogs had not even barked. A 24-hour watch was kept for a fortnight and, on the last morning, several ewes were found to have the mark. The police were again called in. Meanwhile, feelings were running high among the peons, and they watched each other with suspicion. The detectives from Río Gallegos noticed that some of the ewes had been cut in the left ear, and they began to suspect that the culprit was a dog, or even one of the ewes themselves. The remaining 1,300 were divided into nine lots, each of which was penned separately overnight. After an apparently uneventful watch, sixteen ewes of one lot were discovered with the infamous mark. These sixteen were parted off and the remaining ewes were divided into four pens. After a process of lengthy elimination, an old ewe was seen to bite a fore-quarter clean out of another's ear. Her bag now totalled almost 900, and she was duly carried off to the butcher's hut to be executed for her misdemeanours. The contents of the stomach were examined and pieces of half-digested ear were proof enough of her depraved appetite, or perhaps of her wisdom, for she had discovered that the ears of her flock-mates contained a small but succulent amount of mineral salts.

Despite the early misunderstanding, the four Spaniards to the north were to become, by Patagonian standards, frequent visitors to Hill Station and the Cape. José Montes, a native of Oviedo, had arrived in Punta Arenas in 1874 at the age of seventeen and, after working for several months as a clerk in the offices of Bráun y Blanchard, he purchased some horses and dedicated himself to trading with the Indians between Punta Arenas and the Río Coyle. This proved so successful that he bought a schooner in Punta Arenas and expanded his business as far north as Puerto Santa Cruz, providing both Governor Moyano at Los Misioneros and the sub-prefecture at Río Gallegos with basic necessities. Meanwhile he persuaded his brother Pedro and young nephew Eugenio Fernández to join him from Spain. The latter proved to be something of a prodigy among the pioneers of Patagonia. Born in Asturias in 1870, he had originally intended to be a priest but dropped his studies "on the day the village carpenter gave him a book about Magellan". At the age of fourteen, he set sail from Bordeaux to Buenos Aires. The journey took over forty days without a single stop, and the young Spaniard not only threatened to throw the captain overboard if the food did not improve but became so disenchanted with the sea that he seriously considered travelling from Buenos Aires to Punta Arenas on horseback. He sensibly changed his mind and made the journey by ship.

After working for several months in his uncle's business house, writing home to his parents to reassure them that he "was not being too much of a nuisance and that they were not to worry about the Indians", he came into contact with prospectors involved in the gold-rush in Tierra del Fuego and decided to join the hundred or so men who thought they would have better luck at Cabo Vírgenes on the mainland. Collecting half a kilo of gold in two months, he was able to buy 180 sheep from José Menéndez at San Gregorio and joined his uncles and Victoriano Rivera on their land between the Ríos Coyle and Gallegos. As José Montes was largely involved in trading with the Tehuelches, Pedro Montes in poor health and Rivera more attracted to the bright lights of Punta Arenas, it was the young Eugenio who travelled to Santa Cruz in 1886 to discuss terms with Governor Moyano, not meeting a single human being on the way and taking over a month to make the return journey in the depths of winter. In return for acting as manager of the farm, he received several square leagues of camp for himself and thus became a fully-fledged *estanciero* at the age of sixteen.

By 1888 Victoriano Rivera had split with the remainder of the consortium and was sole lessee of the estancia Bahía. For a time he, William and Jack joined forces, establishing a company with the brandmark "C", but the venture was not a success. The Spaniard and the Scotsmen did not see eye to eye on running a sheepfarm or on the division of profits, and within a year the company was wound up. Perhaps as a result of this experiment, William never again formed a business partnership with anyone outside his family. Rivera was an extravagant man who preferred to escape life on the Patagonian pampa. He used the costliest fixtures and soon ran into debt, though it was not to be until 1918 that he was declared bankrupt and sold Bahía to the Rudds. A souvenir of his extravagance still stands in the shape of an ornate house half a mile from the Cape on the track to Hill Station. More a mansion than a farmhouse, this folly was built by Rivera at the beginning of the First World War, and it is said that when he discovered that there was no view of the town from the second storey, he added a third. No one, either Rudd or Rivera, ever lived there.

One result of Rivera's visit was beneficial to the Scotsmen. They purchased two "oxen" from him, in fact the calves of milking-cows. As with all newly arrived animals at Hill Station, they were at once given names: "Bill was almost entirely black, Jack was white with black spotches." They were immediately put to use. In the two previous years, the bales of wool had been rolled down the barrancas to the beach. Now a wagon was constructed with a wooden axle and wheels, and the bales were hauled to the beach by the "oxen" and loaded directly on to the wool-ship. This saved time and also provided a source of amusement, for the settlers had soon learned to make the most of anything that happened out of the ordinary. The animals performed their task so efficiently that the Rudds laid a wager with the Hallidays that the bullocks could go down to the ship and return to the shearing-shed without human assistance. "We loaded up the cart and away went Bill and Jack down the barrancas to the beach. In a short time, they were back at the shed, pulling an empty cart." William had stumbled on one small way of assisting his limited labour force.

62

63

64

65

66

57

8

13 *Death of a Son*

STILL WITH NO help from outside, the Hallidays and Rudds persevered with the slow but steady expansion of their flocks. Early in 1889 they waited with great excitement for the arrival of the *Martha Gale*, for William had arranged with Captain Jones to transport a consignment of 1,000 ewes from the Falkland Islands. He knew the risk he was taking but did not want to repeat the long journey to the Straits. Judging by an entry in Willie's diary for 14 February 1889, the result was a disaster: "Did expect one thousand sheep from the Falklands. When we dipped and counted them, there were 387. The others had died on the voyage." William vowed never to try this again. It was a sad loss, but worse was to follow.

Although both families were now running their own farms, they helped each other out as often as possible. The Hallidays were often "down at the Cape", the Rudds "up at Hill Station". Life was still a family affair, and in 1890 Patagonia dealt that family its bitterest blow. On the afternoon of 17 January the 12-year-old Willie went down to the Cape to help with the shepherding. Margaret Rudd later told Mabel: "The first time I remember seeing you was when Willie asked Mother to lift you up on horseback to say goodbye before he left never to return." After work, Willie asked his Aunt Anne to grant him the favour of riding back to Hill Station on Uncle Jack's horse Picasso, recently purchased from the Indians. Jack was across the estuary at the time, getting provisions from a visiting ship, and Anne insisted that Willie wait till he returned in the evening. Permission was duly granted, but as it was a dark, stormy night with no moon, Willie decided to wait till morning. Before going to bed, he caught and tethered Picasso and wrote his diary: "Uncle Jack did come and feched over with him two bags of flower and 3 tins of treckel and 1 case of brandy". It was to be the last entry.

Mabel later recounted the events as they were told to her:

In the morning, Jack got up, made coffee, then pulled wee Willie's big toe. "That you, Uncle Jack?" Willie asked. They drank their coffee together. Willie went to fetch the horse, Jack to fetch the riding-gear. As Willie was leading Picasso to the springs below the house to let him drink before the ride, the horse that Jack had ridden the day before, an old grey nag called Sam, suddenly neighed from the side of Sugar Loaf Hill and cantered away into the distance. Picasso instinctively lifted his head, pricked up his ears and made as if to follow. Willie just as instinctively put his foot on the looped end of the horse's rein, but the loop caught round his ankle. The horse dragged him over the rocky ground. Uncle Jack immediately let the dogs loose and an old bitch at last jumped up and grabbed the horse by the nose. Picasso turned and stood quiet. He had dragged Willie nearly half a league. Aunt Anne was combing her hair at the time and, when she heard the screaming, she looked out of the window, saw what was happening and rushed into the children's room. They all ran in their nightclothes towards the springs. Uncle Jack was beckoning with his hat. The children screamed, "Run! It is Willie!" But they arrived too late. Willie died in his Aunt Anne's arms. While the family carried the body back to the house, Uncle Jack galloped to Hill Station on Picasso to break the news to Mother and Father. But it was Margaret who galloped with him back to the Cape, for Mother was expecting a child and she rode those long few miles at walking pace, led by a silent Father.

The death had a devastating effect on the morale of the families and left the farms seriously undermanned. The child that Mary Halliday was expecting was stillborn during the following March. The shock so affected Anne Rudd's milk that her baby, Thomas Rae, only three months old at the time of the tragedy, suffered serious convulsions for the rest of his short life: "A beautiful boy with soft dark eyes and wavy hair, from that day on his brain did not grow up." He died nine years later on 26 January 1899.

Willie's cortège was a small one and a simple burial service took place on 20 January 1890 in what was to become the family graveyard at the foot of Skunk Hill a few hundred yards behind the house. William knew as ever that he and his family would have to call on the same courage that had seen them through difficult days. They would have to take to heart the words of a song that Grandpa McCall used to sing when the children were tired and hungry, and the wind outside was strong:

"Don't give way to foolish sorrow
Let this keep you in good cheer.
Brighter days may come tomorrow
If you try and persevere.
Darkest days will have a dawning
Though the sky be overcast.
Longest lanes must have a turning
So the tide will turn at last."

THE HALLIDAYS and Rudds realized they had sooner or later to employ outside labour. It would have been convenient to turn to the local Indians but, although employed by some farmers at the busiest times of the year, they were mostly considered irresponsible, their apathetic attitude to life and their frequent illness and drunkenness making them unsuited for even the most basic work. Despite their excellent horsemanship, they proved ineffective shepherds, partly because they lacked the necessary patience, partly because, being born hunters, they failed to grasp the meaning of protecting and breeding docile creatures such as sheep. Instead William and Jack employed a hotchpotch of peons of different nationalities.

To find labourers was always a problem for the sheepfarmers of Patagonia. The trouble was that few Argentines from the north chose to suffer the even greater hardships of the far south, and those who did often had dubious reasons for doing so. As early as 1867, the editor of the *Standard* in Buenos Aires wrote: "The greatest drawback here is the scarcity of labourers, and the slovenly way in which the few who do work get through a very small amount." With much of the Argentine pampa being given over to cattle-farming, most peons had little or no experience with sheep, and this was particularly noticeable at shearing-time when some outsiders, such as Akers, were properly shocked at what they witnessed in Patagonian shearing-sheds: "The way they shear sheep would drive an Australian squatter out of his senses. They first lie down the animal, and then proceed to cut it and slash it about as if the objective was not only to take off the wool but to remove at the same time an equally large portion of skin." Fortunately in such a cold climate the wounds tended to heal quickly and there was seldom risk of infection.

At Hill Station, outside shearers were rarely employed in the early years as there were enough experts among the family. At the age of fourteen, Archie was already able to shear 200 sheep in a day. The peons were expected to perform more menial tasks around the camp. Antonio Flores, a Spaniard, turned up one day late in 1890 as if from nowhere, and Antonio Boleto, an ''Austrian'' carpenter, helped put the finishing touches to the houses both at Hill Station and the Cape. They were joined the following year by another Spaniard, known as Pelado or Baldie, who was drowned soon after his arrival while collecting gulls' eggs on the Río Coyle. Another intermittent employee was Carlos Nap, a relatively well educated German who had been serving in the Chilean navy but had now taken to wandering from farm to farm in southern Patagonia. He was to be chiefly remembered for crawling up the barrancas on all fours each Sunday evening after a session at the bar or boliche, built down on the beach in 1895. Some peons were Chilotes from the island of Chiloé on the Pacific coast of Chile, who first sought work in Punta Arenas and then drifted north into Argentine Patagonia. They were considered reliable workers as long as they were kept under supervision and were not faced with emergencies. According to some of the stories they told the Hallidays and Rudds, their place of birth contained its share of Patagonian mystery. There were rumours of a secret sect located in a deep cave whose recruits were expected during their six-year indoctrination to stand under a waterfall for forty days and forty nights to rid themselves of all effects of Christian baptism, to kill a best friend to show they no longer had sentiments and to disinter recently-buried Christian corpses, from the skin of which they made a special waistcoat. Fully-fledged members were said to be able to drive men mad, to spread disease, to divert the course of waters and to change into animals. To the Hallidays' best knowledge, no member of the sect turned up at Hill Station. Other peons included sailors of various European nationalities who found themselves stranded in Río Gallegos, a Scottish shepherd from the Falkland Islands who had no pretensions to becoming a landowner, and men of no fixed abode or country who drifted down off the pampa in search of work and about whom it was unwise to ask questions.

There were to be some problems among this motley band, especially when alcohol was available. Annie Halliday wrote in her diary for 20 May 1900: ''Pinero and Aultman had a fight on the beach, Pinero was going to stab Aultman if he had not been

69

70

stopped." At some neighbouring farms, such as Moy Aike, where the family of settlers was a small one and more outside labourers had to be employed, matters were often worse. Henry Jamieson recorded in a matter-of-fact way the troubles among his peons, and there is a tone of regret in his diary at how much more successful his farm and Patagonia might have been had there been peons "of the right sort", and had there been less drinking.

> *22 September 1897:* Mr C. Johnston, gentleman, does not like to lower himself to cut wood, and for that reason gives in his notice.
> *28 September 1897:* Men all sober today.
> *3 April 1898:* Palm Sunday. Crichton and Pedro Chico had a hen fight. Pedro got the better of it. Crichton lost a lot of blood, and had a black eye.
> *3 May 1898:* Men sweating whisky out of them.
> *18 January 1899:* Urquart came...very obstropulous and would not go until I got another man in his place.
> *25 December 1899:* Blinkhorn shot Arthur Stephen about 5.30 pm.

Particularly difficult jobs to fill were that of foreman or *capataz*, and cook. As far as the other peons were concerned, the latter was the most important man on the farm and was expected to make bread every day, to have a constant supply of coffee, tea or maté at the ready, and to provide a varied diet with a dessert twice a week. At Moy Aike the pressure of work sometimes proved too much:

> *26 December 1894:* Cook drunk and nothing cooked for dinner.
> *9 January 1898:* Carpenter tight, Crichton boozy, cook off work.

Even when sober, the cook was not always a success:

> *8 September 1894:* New cook not showing much culinary capacity until now.
> *10 February 1894:* The cook kissing baby all day.
> *12 November 1896:* Caught M—— kissing the cook at Hallidays.

In fairness to the peon, the work he had to endure in Patagonia was often excessive and the conditions abominable. Whatever his official position on the farm, he was expected to be a jack-of-all-trades, and a man contracted from Buenos Aires as bookkeeper and accountant could find himself being ordered to cut up the meat rations and fill up all his spare time in loading bricks into some mule carts. Though a man could be reasonably certain of finding work during shearing time, he could afterwards be out of a job. Besides, because Patagonia

was so out of touch with the modern world, there was no norm for
wages, and often no contract. As Akers remarked: "They are
absolutely ignorant of the ways affairs have shaped themselves in
Buenos Aires. And they do not believe the employer when he
explains that the articles they wish to purchase are imported and
have to be paid for in gold…every peon considers he is being
cheated." If a Chilean labourer caused any trouble, he was sent back
across the border, as it was not in the interests of the Argentine
authorities to have the area full of illegal immigrants.

The labour problems would eventually come to a head in the early
1920s. Dissident workers formed themselves into Sociedades
Obreras, claiming that "non-Argentines" were becoming rich at the
expense of the peons and were showing little respect for the
Argentine nation. They were led by Antonio Soto, a 23-year-old
Galician who saw Patagonia as a fit place in which to fulfil his
political dreams. Having arrived in Argentina to escape military
service, he drifted into the theatre and the fringes of the anarchist
movement and, arriving in Río Gallegos as a scene-shifter with a
touring company, took to heart the plight of migrant workers,
especially the Chilotes. He was elected secretary-general of the local
workers' union, organized strikes and, with a gang of amateur
revolutionaries, set about throwing the farms into disarray. He
pointed to the immense wealth of such men as Mauricio Bráun, who
in 1902 had imported piecemeal to Punta Arenas a palace from
France. He referred to such men as: "…these egotistic gluttons of
lucre…these twittering pachyderms with their snapping teeth and
castrated consciences" (Bayer).

The estancieros and their supporters retaliated by establishing La
Liga Patriótica, claiming that the workers were inspired by non-
national communists and that, if it was true that the majority of
settlers had been born outside Argentina, it was also true that, had it
not been for their sacrifice, endeavour and foresight, Patagonia
would still be largely an unexplored desert. Buenos Aires, they
claimed, had ignored developments in the south, yet the farmers had
at all times respected the fact that they were on Argentine soil. To
make their point, the members of La Liga Patriótica were given the
national coat-of-arms to display at their estancias and agreed to raise
the national flag in a visible place every Sunday and on national
holidays. The dissident peons, for their part, roamed the camps in
gangs. Shearing-sheds were burned to the ground and fences were

cut, with the result that the flocks mixed, the rams and ewes mated at the wrong times, and lambing and shearing took place too early, too late or not at all. There were accusations and counter-accusations of murder and arson. The settlers had little faith in the government or in the protection offered by the national army and police, so, in the face of continuous strike-action and hostility, they formed their own army, La Brigada, which a farmer was able to call upon in trouble. When the national army finally took action, it murdered 120 peons at what was to become known as ''The Massacre of La Anita''; 300 Chilotes were rounded up at the prize estancia of the Menéndez-Bráun family, after several hostages had been held there. The estancieros selected those suitable for work, and the rest were shot, reputedly into graves they had been forced to dig themselves: ''They went to their deaths with a passivity that was truly astonishing'' (Bayer). One of the hostages is said to have demanded thirty-seven corpses in retaliation for thirty-seven horses that had been stolen. In Patagonian style, the story did not end there: the commanding officer, Lieut. Héctor Varela, was assassinated in Buenos Aires, the assassin was in turn shot by a cousin of Varela who had had his testicles shot off by the rebels, and he in turn was murdered by a Yugoslav midget in an asylum for the criminally insane.

In however bloody and tragic a way, the peons made their point. New regulations governing the working conditions on Patagonian farms were introduced. To a Scotsman some of these conditions seemed ridiculous. Archie Halliday later recalled that he was issued with a list of food required by the peons which included such items as tinned fruit: ''And what about the calafate berries?'' There was one requirement that neither he nor his peons understood. ''What's that?'' he asked. ''We don't know,'' came the reply, ''but we've got to have it.''

In the 1890s, however, the Hallidays were concerned with more immediate matters. The question of legal ownership of Hill Station still had to be resolved. In December 1893 the Argentine government at last fulfilled the assurances given by Carlos Moyano in 1885. Under what was known as the Grümbein law, vast tracts of land were granted to military leaders at 1,000 pesos per square league, and the original lessees in other parts of Patagonia were given an opportunity to buy their land freehold at the same price. The Hallidays and Rudds were among those settlers who took up the offer. They could now develop their farms with greater

confidence. Metal windmills were erected to draw fresh water as the flocks increased. Fences were erected to divide paddocks, and however laborious and costly this might have been, the Hallidays and Rudds were grateful that they were at least able to hammer the posts into their ground, as on farms nearer the Andes, where the land was far rockier, piles of stones had to be built round each post. Any expansion, however, was always at the mercy of a natural calamity, especially of a bad winter. In 1899 the Hallidays lost almost 3,000 sheep in a single winter and the horses became so hungry that they were found eating the wool off dead sheeps' backs. There were also frequent outbreaks of scab, necessitating dipping, which was costly and time-consuming. There were also accidents: in 1908 the shearing-shed at Hill Station was burnt to the ground full of recently-clipped wool, and in 1913 the *Coyle* sank with the Rudds' wool on board. And Willie's death had made the Hallidays and Rudds aware of life's day-to-day risks:

> *26 November 1903:* John had kick in face from horse, breaking cheekbone.
> *24 May 1905:* Archie going to Buenos Aires to see about his foot, badly crushed by horse.

Moy Aike continued with its share of accidents:

> *2 December 1897:* Came home. Found Charlie Johnson had been dragged to death by his horse.
> *3 December:* Went up to see C. Johnson, stretched the tent over him; making a coffin.
> *5 December:* Doctor Fenton and el juez came and held an investigation on C. Johnston. Took him down to the settlement. Buried him at 5.20 pm.

The Patagonian sheepfarmers, like those of the Falkland Islands, depended for their prosperity on the world price of wool. Some years there would be a boom, such as during the First World War when they were able to undercut their rivals in Australia and New Zealand, but in other years there would be slumps, such as in 1920 when inflation, new taxes and stricter tariff barriers led to disaster, compounded by the labour unrest.

Much depended, too, on the type of wool in demand. A prolonged search was made to discover the cross-breed most adaptable to the climatic conditions of Patagonia and yet producing the wool most appropriate for the world market. The animals originally purchased from José Menéndez, which themselves came from the Falkland

Islands, proved to yield wool that was not only too coarse but insufficient in quantity to compensate for the few head that the land could sustain. The Hallidays were forced to experiment, and each breed had its disadvantages. The wool fibre of the Cheviot, for example, was weak at the root and on farms such as Hill Station and Cape Fairweather, where there was an abundance of thorny shrub, it was apparent where the flocks had been grazing from the amount of wool discarded. The Lincoln was a prolific producer of wool but so weak in the leg that it easily collapsed and became frozen to the ground, where, if not rescued, it died a lingering death, and in soft camps it was particularly susceptible to foot-rot. The Merino, imported to Patagonia from New Zealand in the late 1880s, was at first used extensively for crossing but was later generally discarded as farmers found it difficult to achieve a constant type, the lambs in each successive generation becoming smaller in body. And the sort of detail which the Patagonian sheepfarmer, out of touch with the increasing knowledge of his Scottish counterpart, discovered for himself was that "very often the wool which covers the face of the Merino gets soaked with poisonous dip which affects the eyesight, inflaming the eyes for a few days; consequently it is necessary to clip this wool, which means a big loss of time" (Jamieson). Just as cattlefarmers to the north came to favour one particular breed or another, becoming "Aberdeen Angus men" or "Hereford men", so different farmers of Patagonia put their faith in different breeds, depending much on the type of their original flocks, the condition of individual camps, judgement of future demand or just personal taste. The Hallidays eventually settled for the Lincoln crossed with their original stock, while some of their nearest neighbours preferred the Romney Marsh.

In order to shield themselves from fluctuations in the price of wool, they tried to go into the frozen mutton business. In 1894 Harry Wood and the brothers Walter and Tom Waldron purchased the vessel *Oneida*, fitted her with refrigerating machinery and anchored her off their estancia Condor. The animals were slaughtered on shore, frozen on board and transferred to ships of the Houlder Line to be taken to Britain. The operation became viable enough for the Houlder Line to open a freezing plant at the Río Seco in conjunction with the ubiquitous Importadora y Exportadora de la Patagonia. A similar freezer was opened in Río Gallegos but was so troubled with discontent among sheepfarmers at the prices offered to them that

76

77

frozen mutton never became a serious commercial proposition.

Despite problems, the sheepfarm of Hill Station expanded and flourished. By 1897, the first year for which records exist, the returns were prosperous:

	£	s.	d.
12 bales	17	0	3
27 bales	220	14	4
65 bales	587	1	5
18 bales	155	6	5
19 bales	155	10	7
1 bale	4	12	8
sheepskins:			
4 bales	23	19	4
5 bales	31	13	6
1 bale	3	1	10
100 wethers	300	0	0
TOTAL	1499	0	4

By 1893 there were eight Halliday children, the last being James, always known as Jimmy, born in 1893. Another child was stillborn in 1895. As the younger children grew up, they played their distinctive roles in the running of the farm. Margaret, or "Queenie", was accepted as the "brains" of the family, writing all her father's business letters and keeping the farm's accounts. Anne was the most domestic of the girls with the nickname "Cookie" and seemed older than Margaret because she was less shy. John was the most serious of the boys and was destined to take over the running of the farm. Archie, the best horseman of the family, was considered something of a "gadabout". The youngest children took it in turns to rise early to light the fire for the cooker, which this time had been successfully removed from a shipwreck north of the Cape, "a wonderful thing with brass rails all round the edge to keep the pots from sliding off in stormy weather". Mabel recalled once being found fast asleep in front of an unlit fire. When five years old, she was allowed as a special treat to carry the tea from the house to the shearers in the shearing-shed: "Halfway across the cañadón there was a large calafate bush, known as 'Mary's bush', where whoever was carrying the small wooden barrel full of tea was met by Mary. On this occasion I tripped over when reaching the bush, and spilt the tea. I received a scalding and a scolding and was made to return to the house. There I refilled the barrel and this time carried it all the way to the shearing-sheds — where, amid much teasing and laughter, I was made to explain why the tea was late."

Physically, the Hallidays were remembered by their contemporaries for their height and for their eyes, bright blue, piercing and some with a "sleeping" left eye. Lucas Bridges, who visited Hill Station on several occasions, recorded: "It was said that if laid out end to end with their shoes off, the Hallidays would have totalled a length of forty-eight feet." And he gave this description of Archie:

> When he straightened himself up, Archie was over six feet tall, but he had faced so many gales for hours on end that, like most of us, he walked with a stoop. Lean and wiry he was, as behoved a man who led an active outdoor life, but what struck me most about him was the quizzical yet kindly humour of his eyes.

The Hallidays were known as excellent horsemen, and anyone who married into the family was expected to possess the same quality. Henry Jamieson, who married Mary in 1897, had taken part in the mammoth drive from the Río Negro. Margaret Sinclair, who married Archie in 1913, had at the age of seventeen ridden the return journey from Puerto Santa Cruz to Isla Doble near San Martín de los Andes, a distance of "over 170 leagues". Her gear consisted of a bit, bridle, reins and sheepskin saddle: "Had not the trifling obstacles of the waters of the Lago, the cordillera of the Andes and the glacier O'Higgins stood in her way, I am sure that she would have continued to the shores of the Pacific and so have had the honour of being the first of the 'weaker sex' to ride from the Atlantic coast to the Pacific" (*Standard*). On his death-bed in 1943 Archie organized a "raid" from Hotel Güer Aike just outside Río Gallegos, to Puerto Santa Cruz, a distance of sixty leagues. The Buenos Aires *Standard* announced proudly that "six descendants of Patagonian pioneers of British descent finished among the first dozen riders".

In 1898 the original house at Hill Station was demolished and a larger one built, with a gallery along the eastern and northern sides. An important addition was "the schoolmaster's room". With their new prosperity, Mary and William were determined to give their children a more formal education than they themselves had received.

Their first attempts were not a success. In 1897 the two elder boys, John and Archie, were sent to the Falkland Islands to attend a school established in Port Stanley by the islands' chaplain, who had become Dean Brandon when the new Christ Church Cathedral was consecrated in 1892. It was run by a Mr and Mrs Summers, and Mary and William were confident that it would be an upright establishment. They themselves had been officially married by Brandon when he arrived in the islands in 1877 and they remembered him as a much-respected pastor, visiting even the most remote settlements on an annual three-month journey made by foot and on horseback, during which he acted not only as postman but also as entertainer, somehow managing to find room in his saddle-bags for a magic-lantern. He had established Sunday schools and a library, fought alcoholism among the islanders and helped to publish both the Falkland Islands Magazine and a church paper. But in August 1898 Agnes Jane received a letter from Archie which suggested that all was not well at the school:

> I have had the cane about 30 times for being late because Mrs Summers makes me late on purpose. I don't like either Mr or Mrs Summers or Doc Brandon. They make me do all their work. If you are lazy to write to me I will be lazy to write to you. I remain your loving brother, A. Halliday.

Mary decided to take the next boat from Río Gallegos to Punta Arenas and from there to Port Stanley, where she would be able to renew old acquaintances and see for herself what sort of education her sons were receiving. She found them "in rags and acting as scullery-boys", and brought them straight back to Hill Station.

Mary and William decided on a different plan. They arranged with the shippers Spearing & Waldron for a tutor to be sent out from England to live in at Hill Station. In this way they would not only be able to keep a strict eye on their children's progress but would also have another hand about the farm. The first young man to be contracted was one Henry Standhope, who arrived late in 1899. He was not altogether a success. An educated Londoner, he seemed so stylish among the rough hills of Patagonia that he became something of a laughing-stock among the children. He was also suspected of stealing from their collection of Tehuelche arrow-heads: "Mister Standhope has been taking some of my arrow-heads claiming they are his." Finally on 1 June 1900: "Mr Standhope packing. He got the sack four days ago." The reason for his dismissal, however, could

not have been too serious, for a few days later he was down at the Cape: "The children went to school with Mr Standhope for the first time." And the following day: "Old horse Joe ran away with cart with schoolmaster Mr Standhope and broke the cart." The Rudd children, too, were evidently laughing at the tutor. He was followed by Harry Scott, who arrived in April 1901. Son of an English grain-merchant, he had come to Patagonia as much for pleasure as pedagogy and, as a result, was helpful around the farm and popular with the children; he was referred to as "Harry", whereas his predecessor always remainded "Mr Standhope". But Mary and William were still dissatisfied. After a year they dismissed Harry and decided that as soon as they could afford it they would send their younger children "back home" to school in Scotland.

The running of the farm, the expansion of the flocks and the upkeep of the house occupied most of the Hallidays' time. Daffodils and wallflowers were planted in the garden. A kitchen garden was established in the same cañadón in which the family had spent their first months. It needed regular watering and trees were planted round its perimeter as protection against the wind: "and as they would take years to grow, there was not a moment to lose". By 1892 the Hallidays were growing enough potatoes for their own consumption, and later that year "Mother set strawberry plants from the Falklands and planted apple seeds".

There was little time for leisure. The only organized entertainment took place on Sundays. Although there were no Bible-readings or hymns, Mary and William believed that Sunday should be as far as possible a day of rest. Boat-races were held on the estuary, horse-races on the beach and games of polo, cricket and football on the level expanse near the house, with tins of fruit as prizes. Sometimes a picnic would be organized down at the Cape. The children would go in search of Tehuelche arrowheads and scrapers, and arrange them neatly on boards. Fossils could be found by digging into the cliffs above the beach. In general, however, it was still a question of making the most of the ordinary day ordinarily passing. The "bullock-wagon", for example, served as a source of amusement long after its place at shearing-time had been taken by larger carts pulled by horses. Some of the girls once drove it down to the Cape to fetch a small pig to take back to Hill Station: "Mary and Mabel sat in the back keeping an eye on the pig while Agnes Jane took the driver's seat. Alas! The wooden yoke came adrift from the shaft, the bullocks walked on, the cart tipped up, the passengers slipped out, and the wee piggie ran all over the place in its sack.'' On another occasion, Mabel and her mother drove over to Kilik Aike Norte: ''Annie and Archie hid in a large empty tank at the top of the hill and, as the cart approached, started kicking the inside of the tank and made such a din that the bullocks took fright. As did the passengers! But we all went home in the cart and laughed all the way.''

Such small incidents became memorable. An Indian squaw once walked uninvited up to the house and offered the children, of all things, chewing-gum. One of the typical titbits of advice that the Indians gave the settlers was that to keep the teeth healthy and white in Patagonia they had only to chew the gum from the ''incense-bush'' or *maken*, a thorny, scraggy type of juniper. Found in tiny lumps on the underside of the lower branches, the gum was rolled into a pellet, held in front of a fire on the point of a stick and the melting globules, caught in a bowl of cold water, were kneaded until they had the texture of putty. A small ball of *maken* lasted for up to a week in the mouth without diminishing in size or losing its ''most agreeable flavour''.

Mabel recalled another incident. One night late in 1891 the children were woken to watch an eclipse of the moon and rushed to the living-room, from where they could look up at the sky as the roof of the house had not yet been completed. ''Grandpa McCall stood next to me and, with his eyes transfixed on the moon, spat a globule of tobacco right on top of my head.'' She did not pass comment at the time.

In July 1895 an event of greater significance took place — the killing of the last puma ever seen at Hill Station. Three of the children had taken a motherless lamb to graze on a grassy ledge above the barrancas. The old dog Rover suddenly smelt out a puma and chased it into a gap in the rocks. Annie and Margaret hurried to see what was happening but, at the sight of the lion, they were soon running quickly in the direction of the house, shouting vainly for Rover to follow. Margaret was reluctantly persuaded to show her father where the lion was lurking, and he shot it through the head. It rolled down the barrancas on to the beach, where Mary was walking with the young Jimmy. She heard the shot and saw the lion fall not far away from where they were. In the excited chatter that followed in the kitchen, Jimmy, who had only just learned to talk, made it clear that it was he who had killed the lion. Archie, on the other hand, had developed a mysterious stomachache, which the rest of the family were quick to diagnose as "cold feet".

With no companions of their own age apart from their cousins at the Cape, the children attached more than usual importance to animals. As on any good sheepfarm, the sheep were recognized individually, and those with particular characteristics were given names. And, almost daily, there were events of significance in the animal world:

> *9 April 1901:* A fox carried off the gander.
> *16 April 1901:* A fine day...killed the fox.
> *8 December 1901:* On Jim's birthday there was a funny brown bird. We had him sitting on Pollie's perch in the front and Margaret and I dug worms for him. There was a hawk killed a pigeon right near the door.
> *29 April 1901:* Fine day; dipped all the rams; all the geese died but two, and they are sick also.
> *30 April 1901:* The geese were poisoned by drinking the dip water.
> *22 April 1902:* The ibis flew away today. They cry out "Black Jock" and always leave between April 19–22.

As well as dogs, cats, motherless lambs, hens, turkeys and geese, the Hallidays kept some more unusual pets over the years. They once saved an injured calendar-lark or *calandria*, in fact a type of mockingbird, and a multi-coloured bandurria, a long-legged bird resembling an ibis and so-called because its remarkable singing sounded like a bandore, an instrument similar to the guitar. The "wee ostrich" proved to be tame and confident if caught young enough. There was a pet guanaco called Herbert. Even the "cursèd lion" could be as tame as a cat if found as a still-blind cub, reflecting its inherently unaggressive nature towards man. Back in Britain the puma was at one time a fashionable pet, Edmund Kean the actor being among those who kept one close to his side. One settler, Ernest Cattle, had pumas "roaming round his house" and Archie Halliday later kept one called Bonzo which, with its claws filed down, lived at peace with the family's dogs: "It seldom ate raw meat but lived on a diet of milk and rice, and on the one occasion it attacked a hen it was given a good thrashing. When taken for its evening walk, its favourite prank was to run ahead, lie in the bushes and spring on whomsoever passed, an unnerving event for the uninitiated. It enjoyed leaping at lampshades, playing with a poloball and swimming. Unfortunately

81

after only nine months as a pet it was somehow poisoned to death.''
Mabel once kept a pet skunk or *zorillo*. Called ''the huffer'' by early
navigators, ''because when he sets sight on you, he'll stand
vapouring and patting with his forefeet upon the ground'', its foul-
smelling fluid could be recognized from a distance of half a league,
and anyone unfortunate enough to touch the animal was at once
transfused with the smell. Prichard wrote: ''Nobody could touch
him, no horse would carry him; and a gentleman of our
company...had reason to regret the curiosity which led him to kick
the lifeless body of the skunk with his only pair of boots on.''
Mabel, however, found the skunk, or *memphitis patachonica*, ''a
very pretty and rather interesting little animal — once the 'skunk-
bag' has been removed.'' One pet which the children always wanted
to keep but never succeeded in catching was the small mail-clad
armadillo. Widespread in northern Patagonia and on much of the
Argentine pampa, where it was considered excellent eating and its
hard back was used as the base for a popular musical instrument, it
seldom travelled south of the Río Santa Cruz, but on the few
occasions one was seen at Hill Station, the children immediately
gave chase: ''It was the sweetest, most inoffensive creature.'' But
they had little hope of catching it because, despite its small feet, it
ran at great speed, was safe from attack by dogs which could not
take the whole of it in their mouth, and burrowed so quickly that it
would almost disappear before a rider could dismount his horse.
On one occasion Archie managed to get hold of an armadillo's tail,
but learned that the animal had another extraordinary defence: ''It
could discard its tail and grow another one within a few days,
though at least the hard rings served as a smart way for Archie to
keep his cravate in place.''

At the turn of the century, hares began to appear at Hill Station.
Introduced from Europe by Curt Meyer, a German settler at
Rospentek, Río Turbio, who wanted something extra to shoot at,
they had multiplied and scattered. These were real hares. The so-
called Patagonian hare was, typically, not a hare at all and was rarely
seen as far south as the Río Gallegos. It had acquired its name
because it resembled ''an oversize European hare on stilts'' and had
often been mistaken for a hare by early navigators, such as Byron:
''We saw several hares as large as some of our deer.'' It was in fact a
cavy, cousin of the guineapig, though not nearly as prolific.
Described by Darwin as a true friend of the desert, it bred only once a

year and had but one or two offspring, born in such an advanced stage of development that they were able to feed themselves the day after birth.

Rabbits arrived at Hill Station later than hares. Imported to a farm at Mina Rica near Punta Arenas in the 1910s, they had proliferated so quickly that the farmer had to use domestic cats in order to get rid of them, Mabel recalling that at one stage he had over three hundred cats on his farm.

As well as the menagerie of pets and the constant events in the natural world, the Hallidays found much pleasure in music. Grandpa McCall taught Scottish folksongs, and Mary would sing lullabies to send the younger children asleep:

"Do your best for one another
Making life a pleasant dream.
Help a worn and weary brother
Pulling hard against the stream."

In 1897 a geologist from Princeton University, Professor Hatcher, camped for several weeks at Gransy's Springs and some evenings entertained the Hallidays with a tune on the violin. In 1902 Anne was given for her nineteenth birthday a gramophone and a selection of small records with the tunes on one side only and some of their titles engraved by hand: "The Bluebells of Scotland" and "Auld Lang Syne" played by HM Coldstream Guards, "Hey, Donal" sung by Mr Harry Lauder, and "I Want to See the Dear Old Home Again" sung by Mr Louis Breise. In 1909 Mabel wrote: "Pianos and pianolas are very popular in town", and a year later Hill Station had a piano of its own.

Like the Indians, the Hallidays found plenty of excuses for celebration, be it a birthday, a visit of a wool-ship or a successful shearing:

31 July 1902: Agnes Jane's birthday, played the gramophone till 11 o'clock. Danced and danced.

31 December 1913: We had a dance — lancers etc. The shearers came up to first-foot us with accordion and and tins for drums.

A special cause for rejoicing was the wedding of one of the Halliday or Rudd children. In 1897 Mary had married Henry Jamieson, the wedding itself taking place at Moy Aike, the reception at Hill Station. On a remote estancia in Patagonia there were likely to be special problems at a marriage:

84

85

20 July 1897: Mary's birthday. Was to have been married but juez did not come. I went for him at night and arrived back at 2 am.

The following day Jamieson wrote: "Married at midday; had fine evening's enjoyment, danced until daylight the next morning." The newly wed Scotsman felt it necessary to point out to a magazine in Buenos Aires that there were a few public misconceptions about Patagonian marriages:

> Allow me to call your attention to a mistake in your journal in regard to the securing of wives in Patagonia; they are not caught with the lasso as stated in your journal but with the bolas, as the feminine fraternity of the Falklands used to catch geese in former times; after being secured with the balls they are led with lasso to the wifehunter's hut and there held until ransom is paid by her friends; should her friends not pay he simply keeps her until he can catch another that pleases him better. Trusting you will not make these mistakes very often, I am, dear sir, yours, a Patagonian.

Another cause for celebration was a visit from a neighbour, who often lived several leagues away but was likely to travel that distance to help out a friend or merely to have a drink. Since 1890, the Hallidays' nearest neighbours had been the Feltons, whose property, Kilik Aike Norte, adjoined Hill Station to the west. Like William Halliday, Herbert Felton had lived in the Falkland Islands, where his father was in command of the small contingent of British marines stationed there. In 1886 he transported 800 sheep aboard the *Rippling Wave* from Darwin to Beckett's Harbour in the Straits, and leaving the flocks in charge of his brother George, who had arrived earlier that year as foreman in the Mallman Company engaged in goldwashing at Cabo de los Virgenes, he rode north in search of suitable camps. He selected Otern Aike, to the west of Herman Eberhart's land at Chymen Aike, and camped for several months with two peons, building a house, sheep-pens and shearing-shed, and then returned to the Falklands to fetch his wife and daughter. They rode from Punta Arenas to the Río Gallegos through the very thick of the bad winter of 1887. After living for three years at Otern Aike, Herbert decided that he preferred the land on the north bank of the estuary and he exchanged his property for Kilik Aike Norte, owned but still unstocked by Eberhart. His brother was sent ahead to build a house: "Landing the building-materials was risky and difficult because of the winds and tides, but at last they were thrown on to the beach, admittedly a few hundred yards from the spot originally intended." Though generally friendly to the Hallidays, the Feltons were richer, decidedly English rather than Scottish and related by marriage to the governor of Santa Cruz. Their only son, Carlos, was born in 1900.

Other neighbours were the Hamiltons at Punta Loyola, opposite the Cape. John Hamilton, born in Wick in 1860, had arrived in the Falkland Islands in 1880 but in 1885 moved to Punta Arenas and took part in the arreo from the Río Negro in 1886–8. He subsequently formed the company of Hamilton & Saunders with the brothers William and Thomas, and owned extensive property on the Straits as well as at Punta Loyola. He married Olivia Heap. One of his administrators, William Dickie, became a good friend of the Hallidays. Born in Old Medrun near Aberdeen in 1861, he was destined to become a Protestant priest but in 1885 moved to the Falkland Islands under contract with FIC. On the death of his wife Elisabeth Murray in 1899 he moved to Río Gallegos, where he became administrator under Hamilton & Saunders and worked in the first freezing-plant in the town. In 1903 he decided to move to the interior but, finding that land on the Chilean border was too expensive, settled in Lago Argentino, founding the estancia Bon Accord at the eastern end of the lake, building the first roads in the area and a ferry to carry passengers across the Río Santa Cruz. He was to suffer the bad winter of 1909 and the following year his entire flocks had scab. But he persevered, until his death at Bon Accord in 1923.

By the late 1890s, the Hallidays were seldom short of company. Many other people of European descent had arrived in Patagonia, some of whom called at Hill Station. The Hallidays would have been the first to confess that their own story was only typical of numerous settlers who, in one way or another, helped to modernize Patagonia, and they would have considered it unjust if in the course of the story of Hill Station at least some of those other settlers were not mentioned (see Appendix). As Mabel Halliday said: "Don't forget Patagonia is very big, and the Hallidays are not the only pebbles on the beach."

The unwritten law of hospitality was at times stretched to its limits, especially when visitors were unable to cross the estuary because of bad weather:

> 23 January 1901: Blowy day. Could not get to Gallegos, so all went for an ostrich-hunt instead.

In 1895 William suggested to a businessman from the town that it would be in both their interests to open a boliche or small café on the beach below Hill Station. This became successful and, a few years later, a small hotel was added. The Hallidays themselves crossed over to Río Gallegos more and more frequently. Since being declared a sub-prefecture, the town had gradually acquired all the trappings of officialdom: a police station, gaol, tax offices, business houses and hotels. But because so many of the settlers in the area were British or English-speaking Europeans it was for many years like a British outpost. As in Punta Arenas, English was the accepted language and gold or pound sterling the usual currency. The German engineer Carlos Siewert, who visited the town in 1894, wrote: "One hears almost exclusively English voices; you get the impression that you've arrived in 'Old England' or at least in the Falklands; apart from the port officials, everything is British — money, sheep, language, drink, 'ladies and gentlemen'." The official Argentine presence and the Spanish language became more prominent after 1898 when Río Gallegos was chosen as capital of the territory of Santa Cruz in preference to Puerto Santa Cruz, partly because its harbour was larger, partly because it was closer to Punta Arenas, still by far the most important town in the area. In 1898, Río Gallegos had a resident population of under 200. Five years later it had trebled, with more and more government offices, a hospital, banks and both a state school and one run by Salesian priests. It soon became the commercial and social hub of Argentina's southern Patagonia and, as business-houses opened up, settlers no longer had to rely on visiting ships or a passing trader for provisions, but were able to acquire a far greater variety of goods in the town itself. More and more social events took place. Almost every nearby European settler would be automatically invited to a baptism or wedding:

14 December 1895: Mabel was baptized in Gallegos by Parson Williams from Punta Arenas.
27 November 1898: Very warm day; we are all going to Gallegos but mother to Mrs Smith's wedding. We left about two in the afternoon. They got married at five o'clock.

A British Club was formed. Each year settlers of Scottish descent held a St Andrew's Day dinner on 30 November, and the chief of police, always an honoured guest, gave the Scots the freedom of the town for the night: "A Scottish shepherd was supposed to play the pipes but often he got tipsy and incapacitated before the dinner and it was too much trouble to keep him locked up all the day.'' A Sociedad Rural was formed which each year held an *Exposición* at which farmers showed off their prize livestock and shearing competitions took place. During the summer months, until the drawing in of the days, horse-races took place every Sunday at a race-track just outside the town on the road to Güer Aike: ''Hardly conducted according to English Jockey Club rules, but nevertheless an amusing function to watch.'' An annual football match was held against a Punta Arenas team who wore blue jerseys with a Union Jack across the front. There were other special events. In 1899 President Roca visited the town, the Hallidays being among those to pay their respects, and in the same year a bullfight took place.

In January 1905 there was particular excitement in town when a gang, posing as Yankee stockmen from Chubut, rode in dressed as Western outlaws and firing sixshooters in the air. Everyone was most amused until it was discovered that the gang had robbed the Bank of Tarapaca and London, shooting the glass insulators off the newly established telephone line as they rode out of town. It transpired, though it was never proved, that the culprits were Robert Leroy Parker and Harry Longabaugh, alias Butch Cassidy and the Sundance Kid, with their pretty gun-moll Etta Place. A few years later, the same gang reportedly murdered a Welsh storekeeper in the foothills of the Andes, kidnapped a rich Argentine landowner and were finally killed by the police at the Río Pico, though no one could be sure of their real identity.

And there was always the excuse for crossing over to the town for the proverbial urgent business:

> I went down to the port of Gallegos
> On business, very urgent, you may surmise,
> But wherever I went I would always hear
> "Come on! Let's shake the dice."
> I trotted around for a fortnight.
> I was never late to rise.
> But business! Not a blessed thing
> Though I tell you I did rattle those dice.
>
> (Madsen)

Trips were made to San Julián, to farms across the border in Chile, to Punta Arenas and even back to the Falkland Islands. But, however appealing the attractions of such journeys and the bright lights of Río Gallegos itself, the Hallidays were always glad to return to the unsophisticated but pleasant pattern of life at Hill Station and the Cape, to that ordinary day ordinarily passing:

16 January 1898: Very warm day. Capt. Brazier and the two boys went shooting. Pressing the last wool. Made cakes. Maggie and Mabel washed. Two carts passed by. Boys came back early, only shot two guanaco and one fox.

6 August 1898: John is out at Los Pozos dipping this last week; he says the sheep are looking splendid and fat. The hens are laying well; we haven't killed any of the turkeys; they are savage beasts and chase the children. Anne is scrubbing the passage and there is a lot of talking going on. Mary writing; Polly is down in the garden and Agnes Jane is washing the dishes.

3 March 1902: Picnic at Cape. Played cricket and rounders. Aunt Anne has nice coach for four. John away to Markatch Aike to buy rams. Windy dry summer, sheep poor.

10 February 1903: Aggie Rudd calls Jim "Little Postman" — he rides down to Cape with mail.

If there was nothing particular to talk about in front of the evening fire, the weather offered itself as a perennial subject of interest and conjecture:

6 August 1898: The weather here is something splendid this year, in fact we do not realize that it is winter yet; there is no snow, frost or cold, sunshine all day but perhaps it will become different with the next moon.

23 March 1906: The heaviest gale of wind I ever remember. My birthday.

Grandpa McCall had his own set of rules for forecasting the next day's weather:

A mackerel sky means not long dry.
A ring round the moon brings rain.
When clouds look swept with a mighty broom
There's a gale or hurricane.
When old Sol sets in a ruddy glow
He heralds a day of sunlight.
But if skies were red when he gets out of bed
Stormy weather prevails before night.

Quietly, at the back of everyone's minds, however, were the creatures they had come to farm and to breed, and upon which their future prosperity or decline depended:

86

87

The day comes and says: Do you hear
The slow water, the water,
The water
Over Patagonia?
And I answer, "Yes, sir, I'm listening."
The day comes and says: "A wild sheep
Far away, in the region, licks the frozen colour
Of a stone."

(Neruda, "El Corazón Magallánico (1519)")

IT SAYS much for the Hallidays' success that, twenty years after their arrival at Hill Station nearly penniless and without possessions, they were able to return to Scotland. The principal aim of the journey was to find suitable schools for the four younger children to attend. It was also the first time for more than thirty years that Mary and William were to see their homeland, and it had even entered their minds that they might retire there for good. Leaving John in charge of the farm, they set out from Hill Station in April 1903, taking with them Margaret, Annie, Mabel and Jimmy. Mabel later described the voyage in her school magazine:

> Before leaving home, we had much packing to do, and a boat to engage to take us over the river, and our passage to bark on the steamer. We left on April 15th and spent three days in Gallegos waiting for the steamer to arrive. The *Río Gallegos* came in on the 17th and a boat was sent to shore, the steamer being too big to enter the harbour itself. We left Gallegos at twelve at night. Next morning we were at Punta Delgada, a place near the Straits of Magellan. We had to go very slowly up the Straits, it being dark with fog. We arrived at Punta Arenas about six pm.

There they boarded the *Victoria*, a vessel of the Pacific Steam Navigation Company, and on 12 May crossed the Equator, William playing the same joke on his children that had been played on him so long ago, laying a hair across the lens of a telescope and assuring them it was the line of the Equator. There were other settlers on board, including Lucas Bridges, who later described his fellow passengers as jovial people "who threw dice to decide who should pay for the drinks; in spite of these wild ways, I grew to like them on the voyage". On one occasion he dressed up a dummy and placed it in Mabel's bed. On 23 May the *Victoria* stopped at Vigo in Spain, and three days later she arrived in Liverpool. William's original journey to the South Atlantic had taken almost three months, his return had taken twenty-eight days.

The Hallidays rented Fairfield House on the outskirts of Dumfries. It had already been arranged that Jimmy should attend Dumfries Academy and he was to spend the next five years there, eventually returning to Argentina in 1909 to finish his education in Buenos Aires, at St George's College. Margaret, Annie and Mabel were installed at Aberdour House School in Dumfries, under the care of Misses Mond and Wright. It was at first strange to be thrown from the remoteness of Río Gallegos into a Scottish provincial city, but the girls soon enjoyed their role as "the lassies from Patagonia" and were called upon to tell many a tale in the debating-hall and in the school magazine *New Leaves:*

> One moon-lit evening a cyclist was riding a lonely road in the southern part of Patagonia. As he rode enjoying the beauties of the evening, he suddenly became aware of a stealthy heavy tread on the road behind him. Turning round he was scared to find himself looking into the glaring eyes of a large lion. The puzzled animal acted very strangely, now raising his head, now lowering it, and all the time sniffing the air in a most perplexed manner. This was a surprise for the lion; he could not make out what kind of animal it was that could roll, walk and sit still at the same time, an animal with a red eye on each side and a brighter one in front. He hesitated to pounce on such an outlandish thing, whose hood smelled so disagreeably, but he kept steadily following until the poor rider was finally so exhausted from terror and exertion that he decided to face the conflict with the lion. Gradually slowing down, he jumped from his wheels and, turning round, thrust the brilliant light full into the face of the lion. This was too much for the beast. The sight broke the lion's nerve, for at this fresh evidence of mystery on the part of the strange riding animal, who broke himself into halves and then cast its big eye in any direction he pleased, the monarch of the forest turned and ran off into the bush, evidently glad to escape from such an awful being; thereupon the cyclist, with returning strength and much devout blessing in his lamp, pedalled home.

Mabel also made the most of her chance in a school debate to show off her prowess in keeping pets: "I have a hawk, a mouse, and a skunk, which can be put through an operation. There are young lions, an ostrich and guanaco — you will have seen them in the zoo — and what would a house be without a cat and a horse? My favourite pet is a parrot. The last ostrich we had feeding out in the garden. A strange dog came in and that was the end of the poor wee fellow.''

William was evidently pleased with Aberdour House, presenting several Patagoniana to the school museum, including a most beautiful condor's skin, which from tip to tip of its wings measured nearly nine feet. During the holidays, the family toured Dumfriesshire and the Highlands:

12 July 1903: We all walked to the top of Ben Nevis.

But Mary and William saw little future for themselves or their children in Scotland. The Gillespie cab business had long been sold, land was expensive and "there were too many people". Above all, Patagonia began to call. Within three months of arriving at Dumfries, Mary and William were on their way back to Hill Station, and nine months later their three daughters followed them:

10 October 1904: We three girls arrived home from Scotland. Little dog Frisky still alive.

They had, after all, eaten the berries of the calafate and were, according to Patagonian lore, bound to return.

Hill Station and Cape Fairweather crept reluctantly into the twentieth century. Wells were bored for extra fresh water, more windmills were erected, and, in 1910, power-shearing was at last introduced. The kitchen garden was extended: in May 1912, Mabel wrote to a schoolfriend in Scotland, "I am writing this to ask you to send out a packet of cabbage seed to Daddy in a paper as he would like to have them set early. You must be very careful that the port authorities don't notice you are doing it or they will stop you." By 1909 William, 64 years old, had grown tired of the hard work of running the farm, and John was appointed manager and a more elaborate house was built next to the original one to accommodate him and his wife Catherine, née Johnston, whom he had married the previous year. The flocks increased and, with the new profits and with the fear of over-grazing the pastures of Hill Station, the Hallidays bought two new farms: Cañadón del Rancho, near Puerto Santa Cruz, of which Archie was appointed manager, and Moy Aike Grande on the Río Coyle, adjoining the Jamiesons' farm, of which Agnes Jane's husband George McGeorge was manager. In 1912 a bridge was built across the Río Gallegos at Güer Aike with the result that people could reach the town directly by horse or car and visitors to Hill Station became rarer. There were still relatively few vehicles in Patagonia. Iván Noya had shipped a small Schacht to Río Gallegos

88

89

90

91

92

in 1908 but it was not till March 1918 that the first car drove into Hill Station, with Lucas Bridges at the wheel: "It was an Albion with a ten-ton truck chassis converted into a touring-car." The journey over the last few leagues had been so hazardous that Lucas threatened never to return if the road were not improved: "We spent an exhausting week cutting the bushes to make a wider track." By 1924 there were still under 600 vehicles in the territory of Santa Cruz, the reason being that, except in the immediate vicinity of the towns, the roads were mere mud-tracks and it was too often a question of the driver getting his shoulder to a ditched Model T Ford and heaving it on to the road again. The horse for a long time to come remained a more practical mode of transport in Patagonia.

Even as the farms prospered, there were some sad and difficult times. Grandpa McCall had died in July 1901, and "Uncle James" Fell, husband of Agnes McCall, also died after catching a heavy cold going to the funeral. One of the Fells' children was drowned at their home at North Arm in Chile. In April 1910 news arrived that Grandmother Margaret Gillespie had died of congestion of the lungs and heart trouble in Dumfries, and was buried next to her husband in St Michael's graveyard, opposite the mausoleum of the poet Robert Burns. And in May 1917, William Halliday fell gravely ill at Hill Station. Archie received a telegram in Puerto Santa Cruz saying that his father was close to death and wished to see him. Characteristically he decided to ride to Hill Station there and then, using the horses he had ridden into town rather than returning to Cañadón del Rancho to fetch fresh ones. Sending a cable ahead to the Reynards at Cañadón de Las Vacas, eight leagues along the track, he set off at a gallop. It was a stormy night, with a south-west gale of sleet blowing head-on. The horse he was riding began to show signs of failing and, just as the saddle was being changed on to the second animal, the first one collapsed. It was the first and only time that a Halliday rode a horse to death. Archie stopped briefly at Cañadón de Las Vacas, was given fresh horses, rode on to the Jamiesons at Moy Aike Chico, drank several cups of black coffee, and reached Hill Station in mid-afternoon. They had to help him walk the last few yards to the house. He had covered almost 200 miles in twenty-four hours.

William recovered sufficiently to travel by steamer to receive treatment at the British Hospital in Buenos Aires. Mary accompanied him and it was, tragically, she who died suddenly at the hospital on 17 June 1917 at the age of 63. As if inseparable, William died ten days later. He was 72. Both were buried in the British Cemetery in Buenos Aires. Many settlers of Patagonia were present at the funeral, including Frederick Haddock, James Roy, William Saunders, Hugo Lively, John MacKae, Dr Victor Fenton, William and Robert Patterson, Alejandro Menéndez Behety, John Hamilton and Herbert Felton. These people knew what Mary and William Halliday had endured, and they knew what they had achieved. It was appropriate that the expression ''an old Halliday'' had become synonymous with an early settler of southern Patagonia.

The original site of the farm at Hill Station was also dying. In 1923 the larger and more modern of the two houses was dismantled and moved to Los Pozos, a few hundred yards above the lake at Kippern Aike. This became the headquarters of the farm, as it was more sheltered and central, the wool being taken by truck to Río Gallegos rather than being loaded on to wool-ships from the northern beach. The Indians no longer wintered there, the few survivors being given a small track of land west of Río Gallegos, but the majority preferring to exist in shacks on the fringes of the town itself.

By the mid-1920s most of the Halliday and Rudd children had married, and some of them had left Patagonia for good. Agnes Jane had married George McGeorge in 1898. Born in Scotland in 1856, McGeorge had emigrated to the Falkland Islands under contract with FIC. In 1886 he took part in the arreo from Río Negro and eventually settled at Paso del Medio. He then acquired some of the land-rights granted under the Grümbein Law on the south bank of the Río Coyle at Guakenken Aike, ''Place of the Race-course'', now known as Cancha Distante. Known as a man who enjoyed company and the bottle, he paid for the building of the British Club in Río Gallegos. They had three children, Janet, James and Mary. Agnes Jane died in Río Gallegos in 1938, at the age of 62, and was buried at Hill Station.

Mary and Henry Jamieson had three children, Helen, Alec and Harry. Mary died in Buenos Aires in 1953 at the age of 77 and was buried at Moy Aike. John married Catherine Johnstone of Dumfries in 1908, died on 17 April 1939 at the estancia El Zorro and was buried at Hill Station.

Margaret married William Bodden, a Canadian, on 22 October 1919 and returned with him to Vancouver, where she died on 21 December 1953.

Annie married Robert Johnston of Dumfries on 8 April 1920 and moved back with him to Scotland, dying at Maxwelltown in 1957.

Archie married Margaret Calander Sinclair on 14 August 1916, and, on her death, her sister Helen "Tillie" Sinclair in 1938. He died on 14 April 1943 in the British Hospital in Buenos Aires, and was buried in the British Cemetery.

Mabel married Ernest "Sandy" Alexander Miller on 14 July 1917. He had been working for the Meteorological Office on South Georgia and Mabel first met him when he was on leave in Río Gallegos in 1910. They were married in Buenos Aires and spent their honeymoon on the *José Menéndez*, bound for Río Gallegos. They had one son, Donald, who decided as a young man that his future lay in Scotland rather than Patagonia. Mabel died on 5 December 1975.

Jimmy married Alice Maud Collins on 5 June 1923, died at Coleta Olivia on 8 December 1966 and was buried at Hill Station.

As for the Rudds, Jack died in 1927, Anne in 1929.

Agnes married William Fell on 15 February 1910 and died on 14 July 1929 at sea, *en route* to Buenos Aires.

Ellen married Ulric Clasen in 1904 and died in Esquel in July 1963.

Mary Jane died unmarried in Exmouth, Devon, on 4 June 1906.

John Rogers died on 27 May 1921, unmarried, in Buenos Aires.

William Gallegos, the first child born at Hill Station on 29 August 1887, in 1918 married Millie Gunn, and died on 9 March 1922 at Cape Fairweather.

Anne Maria, born on 20 August 1891, married Douglas Alexander MacCleod and died in July 1918 in Edinburgh.

Edward took over the running of the farm.

Edgar went to Scotland.

Maud married Carlos Felton from Kilik Aike Norte.

Ethel Bridget married a Kennard.

James moved to Dos Lagunas.

Gerald Ruben moved to El Falso.

Each year the town of Río Gallegos crept further along the southern bank of the estuary. By 1920 it had a population of over 1,000, still small in comparison with the population of Punta Arenas of 20,000. Commerce grew. The Sociedad Importadora y Exportadora de la Patagonia opened larger offices. Agencies were established to represent the interests of sheepfarmers, the Hallidays being represented by "Gallies", founded by a Scotsman, Andrew

Gallie, who, having worked in Punta Arenas, reckoned there was more of an untapped market in Santa Cruz and established his agency in Río Gallegos in 1920. El Banco de Turapaca y Argentina became el Banco Anglo Sud Americano, which expanded into el Banco de Londres y America del Sud. Spanish gradually became the standard language, the Argentine peso the usual currency. In 1923 the first planes flew into Río Gallegos: "We thought they looked like flying silver fish in the sky." The first official Aeroposta Argentina arrived in 1930.

Settlers penetrated to the remotest parts of the province of Santa Cruz, to Río Turbio, Lagos San Martín, Viedma and Buenos Aires, to the very edge of the enormous, advancing glacier, *el ventisquero* O'Higgins. Pioneers seemed to be following the advice of Martín Fierro:

> And whether we keep, or lose our lives,
> It's little that men still care;
> We only need to keep going west,
> And leave it to God to do the rest;
> We'll arrive some day; and, afterwards
> Time enough to find out where.

Ports along the Atlantic coast also saw expansion: Puerto Coig on the Río Coyle, Comandante Luis Piedrabuena on the estuary of Puerto Santa Cruz, Puerto Santa Cruz itself, San Julián, Puerto Deseado and Caleta Olivia. Even the more arid areas to the north were stocked with sheep, however few to the hectare. As if to mock his curse of sterility, an estancia has been called Darwin. By 1937 there were an estimated 7 million sheep in the territory of Santa Cruz, which, at an average of 300 per inhabitant, was probably the highest ratio of sheep to human beings anywhere in the world. Roads were built from Río Gallegos to Lago Argentino and to the Andes themselves, to Sierras de los Baguales, to Mounts Fitzroy, Payne and Stokes. Coal at Río Turbio was mined extensively and a railway built from Río Gallegos. Comodoro Rivadavia in the province of Chubut became the oil capital of Argentina after the substance was discovered in 1907 by two men working for the Dirección de las Minas searching for water.

Murgia
FOTO

A Patagonian Farewell

"No wonder I sigh for the days gone by,
The times that should come no more."
(Hernández)

NOWADAYS it seems that Patagonia has been tamed. Yet the name still intrigues outsiders. It is thought by some people to be a fictitious country like Ruritania or somewhere near the Amazon or next to Mesopotamia or somewhere off the Hebrides, or a disease. It is a place where the old people are made to climb trees, the trees are shaken and, if the poor folk are too weak to hold on, they are hit on the head as being of no more use to society. Patagonia is still the back of beyond, a nowhere place. Jules Verne located two of his novels there, *The Lighthouse at the End of the World* and *The Sons of Captain Grant*. In *Vol de Nuit*, the French author Saint-Exupéry writes of adventurous times spent there delivering mail by air. A "Patagonian sling", presumably a boleadora, was used by the castaways in *Swiss Family Robinson*. The tradition established by Lady Florence Dixie of using Patagonia as a place where children can perform prodigious acts has been maintained in such books as *Patagonian Holiday* by M. J. Ross. In *Last and First Man* by William Stapledon, the human species, almost extinct through cannibalism and disease, escapes to Patagonia and sets up a new civilization under the leadership of an adolescent of prodigious sexual capacity, though, unfortunately, this new society is itself destroyed by an atomic cataclysm. Jaroslav Hašek, creator of *The Good Soldier Svejk*, when he was editor of the family magazine *Animal World* invented a report that packs of wild Scottish collies were terrorizing the population of Patagonia. In James Joyce's *Dubliners* Frank tells Eveline of "the terrible Patagonians" and, at the beginning of Evelyn Waugh's *Scoop*, John Courtenay Boot is putting the finishing touches to a book called

Waste of Time, "a studiously moderate description of some harrowing months among Patagonian Indians". In the late 1970s, two best-selling travelogues, *In Patagonia* by Bruce Chatwin and *The Old Patagonian Express* by Paul Theroux, showed that the word still has popular appeal, even if in the latter book only the last few pages took place in Patagonia, and then only in its northern fringes. When awarded the Booker Prize in 1979 and asked what importance she attached to such literary awards, the English novelist Iris Murdoch replied that they mattered to her as much as if it was raining in Patagonia or not (to which a Halliday might reply that it matters much more than who has won the Booker Prize). Commander W. Campbell, an original member of the BBC programme "Brains' Trust", used the catchphrase: "When I was in Patagonia...." A leader in the *Guardian* newspaper published in London in May 1980 started: "Mr Nixon (in case you've been in Patagonia for the last three weeks) has written a book called *The Real War*...." In 1980 the satirical magazine *Private Eye* published a letter to the Editor which included a reference to "gigantic, muscular Patagonian basket-ball players".

Patagonia is still a place of extremes. Whatever else, man has not been able to subjugate the weather. Modern sailors and naval commanders will vouch for the atrocity of the South Atlantic. The keepers of a lighthouse at the western end of the Straits of Magellan have had to build cement walls to protect their freshwater cistern, some 165 feet above sea level, from waves breaking on rocks below. Stories can still be heard of people finding an 8lb trout outside their back door though the nearest lake or river be eighty miles away, and of a tornado which not only emptied a lake but sucked up the mud at the bottom into stalagmites. Eccentricity still flourishes. It is possible to be greeted with the words, "Hello, I'm from Kent," even though the speaker has never set foot out of Patagonia, and one wealthy settler owns an aeroplane but, having been refused a pilot's licence, drives it round his farm hopping over the fences when no one is looking. Another went bald but managed to grow his hair by thinking of sheep's wool and remained unconstipated by ruminating on sheep's intestines. Suicides, alas, are still not uncommon: "For others the situation has been more tragic, and more than one farmer has been unable to face up to the facts and has taken his own life" (*Buenos Aires Herald*). And some stranger aspects of Patagonia's reputation may not be so far-fetched as at first supposed. Even if not giants, the Tehuelches were the tallest people in modern history apart from the Watutsi of Central Africa. At an average of 6 feet 1 inch, they stood half an inch taller than the tallest men in modern Europe, those of Sutherland, in Scotland. In comparison with the Europeans of sixteenth- and seventeenth-century Europe, they were indeed giants, and in the 1970s it was still possible to see at St Giles Fair in Oxford "A two-headed Patagonian giant, perfectly mummified". Rumours of the existence of prehistoric monsters linger on. In 1922 Martin Sheffield, known as "the yanqui-gaucho", wrote from Lago Epuyen to the National Zoological Museum in La Plata claiming he had sighted a plesiosaurus, a supposedly extinct marine reptile. He requested an expedition to be organized, suggesting that some embalming fluid be brought in case it proved impossible to capture the beast alive. The news was broken to the press and the Patagonian Plesiosaurus became the talking-point of Buenos Aires, even having a tango named after it. One small expedition set out in vain, and the creature soon became poor cousin to the Loch Ness monster. But other grand schemes were planned. In 1939 a German attached to the La Plata museum intended to create "the largest museum in the world", to be sited at el Bosque Petrificado, a fossilized forest at Lago Madre e hija, 170 kilometres west of Puerto Deseado. Over 80 square leagues in size, the forest had been covered by volcanic lava for millions of years and had then gradually been exposed. Frank Mansfeld wanted to build a vast museum full of life-size models of the animals which had roamed over Patagonia in prehistoric times: antarctosaurio, at 50–60 metres long "the largest animal in the world"; smilodon, an ill-named sabre-toothed giant tiger; phororhacos, a hideous bird five metres high. He held exhibitions and press conferences in Buenos Aires and published a lavish prospectus, but the scheme died in embryo. He might have received some compensation from the fact that in 1954 the area was officially declared "Monumento Natural de Bosques Petrificados".

The idea of the existence of an El Dorado has been discarded, but sufficient amounts of gold are still found in southern Patagonia for the story to retain a semblance of possibility. However much one may ridicule reports of people blowing off their noses into the fire and pulling off their toes with their socks, it has been established that this can happen in extreme cases of frostbite. The Tehuelches may not have been totally misled in some of their beliefs: the idea that the souls of the dead ride to heaven on horseback coincides with the fact that the estimated numbers of people who have walked this earth is approximately equal to the amount of stars in the Milky Way; putting their hands to their head at the sight of a new moon is hardly more superstitious than the cristianos' custom of turning a coin, and perhaps they were not so ill-advised in being wary of the written word.

Some modern settlers question at what expense the taming of Patagonia has been achieved. Despite the Welsh successes in Chubut, for example, the original dream has steadily evaporated. In 1910 Prichard was already warning: "The danger takes the form of the dark-eyed Argentine maid who is rather apt to make roast meat of the heart of the Welsh youth." Nowadays about ten per cent of the population are direct descendants of the original colonists and some 5,000 people speak the Welsh language. There are plenty of Welsh surnames but, since Argentine law demands that every individual has a Spanish Christian name, there are Maria-Teresa Joneses and Enrique Williamses and streets called Juan C. Evans. Welsh is no longer taught in the schools. A Welsh newspaper, *Y-Drafod*, is published, but its circulation is dwindling. An annual Eisteddfod is held, but over half of it is sung or spoken in Spanish. Welsh dolls, made in Hong Kong and dressed in Gaiman, are for sale, and there are Welsh tea-shops for the tourists, some of whom believe that the tea served is grown in Wales. There is sporadic singing in the chapels. Linguists from Aberystwyth University arrive by jet to study what is judged to be the purest Welsh in the world. To an outsider, there is something both admirable and desperate in the efforts to keep alive the culture, language and traditions of a small country thousands of miles away, a country which the vast majority of people have never visited and from which each successive generation feels further removed. From the air, the valley of the Chubut looks like a paradise of a sort, an oasis in the middle of cat's-litter; but the pampa is strewn with debris caught in the thorny scrub, there is an aluminium smelting-plant and a Kansas Bowling Alley in Puerto Madryn, and Rawson has an infamous prison for Argentina's political detainees.

There is a strong feeling among settlers of Patagonia that it has never lived up to its potential, that it still does not readily welcome visitors. Although representing twenty-five per cent of Argentina's land-surface, it still contains only two per cent of its population. The poet Madsen once called it "the land of promise under the Southern Cross", but a more common opinion would be: "What a country this could have been" with the afterthought, "It may yet come true." Some sheepfarmers even believe that Patagonia is finished: "Patagonia is dead, not dying.... Today we have a modern town with paved streets, mercury lights, specialty shops, *boîtes* and flood-lit plazas with expensive statues of our country's saviour.... Today we have 25 times as much government as we had 25 years ago but everything else has dwindled to nothing and Patagonia is dead" (*Buenos Aires Herald*). Though rich in oil, Comodoro Rivadavia has never enjoyed the success that was forecast, even when world prices were at their highest; it has often been the pawn in the political games of fast-changing governments, some encouraging foreign investment and expertise, others immediately expelling it; during the oil boom of the 1970s, Argentina is said to have been the only oil-producing country in the world to have made a loss. Even though Punta Arenas is still the largest town in southern Patagonia, it has, like all Chile, suffered from the problems of its country's politics; when Fidel Castro paid an official visit in the 1970s, a plea was broadcast for people not to queue for bread while the Cuban leader was in town. Three small islands in the Straits of Magellan are still the subject of a dispute between Argentina and Chile which has on occasion brought the two countries close to war. Though the islands are unimportant in themselves, the nations' claim to a 200-mile "economic zone" means that whoever owns them can also claim large areas of the South Atlantic, apparently rich in resources.

However difficult their life may be in Patagonia, William Halliday's descendants are grateful for his decision to have left the Falkland Islands. In the 1970s the islands still sent over 2,000 tons of pure wool to the UK each year, but the land had become so barren that it could only support one sheep to every five acres. There had been serious overgrazing, nothing was able to rot properly, chemical fertilizers had little effect in the cold climate, fertility built up by the dung of penguins, seals and geese had almost disappeared, and the wild white clover did not spread because of a lack of bees. As a result of the failure to establish a freezing-plant, and despite talk of the exploitation of kelp, fisheries, offshore oil, berry crops and tourism, the islands continued to depend on wool and on postage stamps, which represented a third of the islands' income. FIC, by then part of the Coalite conglomerate in Britain, continued to own over half of the land and, with its paternalism and the almost feudalistic attitude of many of the farm managers, was accused of sapping individual initiative and responsibility, and of deliberately discouraging the islanders from nursing any possibility of long-term links with Argentina. More importantly, however, the Falkland Islands and their dependencies of South Georgia and the South Orkneys were seen as vital gateways to the Antarctic, under which, it is believed, vast mineral resources lie. It was for this reason, among others, that in 1982 the Falklands became the scene of a short but bloody war between Britain and Argentina.

On the one hand, FIC and the vast majority of the kelpers and contract staff adamantly wish to remain under British control and so retain their ultra-British lifestyle. After all, the islands have been maintained and manned by the British since 1833. They claim to have sent more men per capita to fight in the Second World War than any other British possession, providing five Spitfires for the Allied Forces. They have the highest proportion of Landrovers in the world, and several pubs. They want nothing to do with Argentina.

On the other hand, the Argentine government continues unabated with its vociferous claims to sovereignty. However justified, these have on occasions been used as a convenient distraction from more serious and less glamorous internal issues. The islands are indefatigably referred to as Las Malvinas and a modern law states that any mention of the Falkland Islands in a school or university textbook has not merely to be deleted, but literally cut out.

In the 1970's diplomatic efforts were increased to try to make sense prevail. Whitehall had already begun to consider the cost of upkeep of a colony 8,000 miles away and the logic of the proximity of Argentina, a country with which Britain has had strong and often mutually beneficial links. For a time the dispute seemed slowly to be resolving itself. Argentina built an airport, ran weekly services to and from the mainland, offered scholarships to two of the brightest young kelpers each year to attend a British-style school in the Republic, and rendered its services in dealing with emergency medical cases, and with dry-cleaning. In July 1971, Mabel Halliday wrote from Río Gallegos: "On Friday we were at the Club to a meeting of three men from the Falklands. A happy, pleasant gathering it was, and we are to have a plane a week from the Falklands and people can come over and we go visiting them with no bother of passports. Everyone is pleased." An idea emerged that Britain would cede the islands to Argentina and then lease them back for a transitional period. FIC and most of the kelpers objected.

Whether from diplomatic negligence, from the fact that Britain intended to cut back its defences in the South Atlantic, or from intransigency and opportunism on both sides, the volcano erupted on 2 April 1982 when the Argentines occupied the Falkland Islands or, as they claimed, re-occupied the Malvinas. Although this occupation only lasted ten weeks to the day after the British Task Force set sail from Portsmouth, it caused enough death and destruction to ensure that disputes over the sovereignty to this "fag-end" of the world will continue for decades.

Patagonia, by contrast, seems remarkably peaceful. Across the estuary from Hill Station, the town of Río Gallegos continues to expand. Yet it itself seems caught between two worlds. Modern buildings multiply, as do the shanty towns. The streets are full of modern trucks, and of cars more fit for museums. The British Club survives. The Argentine forces have been strengthened. The annual Rural Show grows in size, but the freezer closed down in 1971. Jets arrive daily from Buenos Aires after a journey of only three and a half hours, yet this fact can paradoxically enhance the sense of isolation felt by people who will probably take longer to return from the airport to their farms. With improved roads and modern buses, the journey to Buenos Aires still takes over fifty hours. The Patagonian pampa seems as monotonous as ever, and the most common cause for accidents on the long road northwards is the driver's falling asleep at the wheel.

As a jet comes in to land at Río Gallegos airport, built on the site of the old race-course and scene of hectic activity during the war, it passes over Cape Fairweather, over Rivera's folly and sometimes directly over Hill Station. The old wooden house is clearly visible from the docks across the estuary, but the port is rarely used these days, the entrance between Cape Fairweather and Loyola Point needs constant dredging, and ships of high tonnage can still only enter, if at all, at high tide. It is almost impossible to persuade anyone to cross the estuary by boat. A visitor must travel by road, by way of Güer Aike, then along a couple of leagues of the main road, down past Kilik Aike Norte, past the turning off to the left to Los Pozos, and finally along the bumpy track to the northern edge of the estuary. The town of Río Gallegos seems so near, yet so far.

Since Mabel's death in December 1975, the house has been

uninhabited and has recently been emptied of its contents. The old "bullock-cart" stands outside the house in the overgrown garden. The door on the privy is off its hinges. The wind beats old ivy against the windows and the rusting corrugated-iron roof. Inside, the wooden floors slant at strange angles. There are no more arrowheads arranged neatly on the walls, no blind hen in the yard, no Mabel ringing on the wind-up telephone to hear the latest gossip from Los Pozos or Cape Fairweather, speaking her English and limited Spanish in pure Dumfries brogue, or listening to the World Service of the BBC which, for some Patagonian reason, is picked up extremely clearly. One peon lives in a hut close by. The only other visitors are vandals. It is just possible to see where the kitchen-garden once stood in the cañadón and the foundations of the boliche and hotel down on the beach. There are no condors or pumas. A guanaco is a rare sight, a "wee ostrich" even rarer. The last pure blood Indian is dead and, because of the constantly shifting soil and the increasing number of marauders, it is almost impossible to find the once-abundant flint-chippings and scrapers. There is no sound of human voice, only the bleating of sheep, the hum of the traffic from across the estuary, the occasional roar of a jet and the squabbling of seagulls. Across the hills at Los Pozos, along the estuary at the Cape, and further north at Moy Aike, Jimmy Halliday, Ramie Rudd and Bill Jamieson continue the long battle against the vagaries of the weather, the vicissitudes of politics and fluctuations in price of wool.

Mabel now lies in the graveyard at the foot of Skunk Hill, with Willie, and with her parents, whose bodies were brought back from Buenos Aires to their proper resting-place. If a visitor listens very carefully, he may hear, from the distant past, the sound of girls giggling, the growl of a puma, the shot of a rifle, the rumble of a cart or of a slow cortège moving towards the graveyard. Or was it the heavy breathing of a giant?

And if he closes his eyes, he may see an old Indian lumbering up the cañadón, or a peon on all fours returning from the boliche, or Mary and William Halliday and their children coming out of the house to be photographed in the sun:

"A laugh and a sigh
A kiss and goodbye is our life.
Is it worth so much fretting?
It is a merry life on the whole.
Courage, comrade." (From Mabel's last letter to the author.)

OTHER CLOSE neighbours of the Hallidays included the brothers Christopher, Peter and John Smith who had moved from the Falkland Islands to settle at Coy Inlet at the mouth of the Río Coyle, and the Reynards at Cañadón de las Vacas, almost exactly halfway between Río Gallegos and Puerto Santa Cruz close to the Atlantic coast. Reynard was the man who accompanied Governor Almeida from Punta Arenas on his diplomatic visit to the Falkland Islands in 1877. He had arrived in Buenos Aires at the age of twenty and, after travelling extensively over the provinces of Santa Fé and Córdoba, in 1874 went from Montevideo to the Straits of Magellan, where he opened a business-house with Elias Bráun. This was thrown into disarray by the mutiny of 1877, but the sheep he imported from the Falklands to Isla Isabel proved a success and he became the first man to sell wool to Britain, the first to establish a *grasería* or grease-plant in Patagonia and the first to import a shearing-machine. But in 1890 land in Magallanes was auctioned by the Chilean government and Reynard was bought out. Hearing that land in Argentina was cheaper, his friend John Hamilton persuaded him to move to Cañadón de Las Vacas, to which, in the company of George Greenwood, he drove 2,500 sheep. Such was the fight against the pumas that Greenwood became demoralized and sold his shares in the farm, leaving Reynard in sole ownership.

Other neighbours were John Redman and William Woodman at Güer Aike, but it was one of their shepherds, Ernest Cattle, who made the more frequent visits to Hill Station. Little was known about Cattle's past except that he was thought to be an officer in the British Navy who had abandoned ship in Punta Arenas. He moved from Güer Aike to settle at Cerro Frias at Lago Argentino, where some hills are still called Sierras Cattle, with another Englishman, Ernest Game, who disappeared mysteriously. In 1902 he transferred

96

97

to Cerro Buenos Aires with "Mister Jack" Harris but, like her, did not take kindly to the arduous life and made frequent trips to Río Gallegos, eventually selling his rights in Cerro Buenos Aires to the Yugoslav brothers José and Jerónimo Stipicic, and retired for good to England.

Another visitor was Charles Henstock, who in 1908 married Emily Felton of Kilik Aike Norte. An Englishman, he had settled at the foot of Mt Comisión on the banks of the Río Centinela which flowed into Lago Argentino. Having suffered abominably in the bad winter of 1901, in 1907 he purchased land on the banks of the Río Calafate. He was known as a fanatic puma-hunter, and later confessed to the Hallidays: "It is true that at the time I was only 14 years of age and the hazardous life I had to lead, looking after my sheep and hunting and killing the countless lions that were lying in wait for a chance to undo the result of my labours, represented to my mind but one of the many sports I had formerly practised in my country." He and his wife finally emigrated to Canada.

A particularly popular visitor was Dr Victor Fenton, who was ready to be called out if any of the family fell ill. Born in Sligo, Ireland, in 1865, he trained as a doctor at Trinity College, Dublin, worked for three years in London, but with a yearning for travel, joined his brothers already working in Patagonia. In 1889 he settled with his brother Thomas at Fenton Station on the Straits, and then moved to Río Gallegos, where his brother Arthur was official doctor to the Governor, Ramon Lista. Arthur was not enjoying the job, so Victor took over. He was a popular man, travelling great distances even in midwinter when news reached him that someone was ill. On one occasion some prisoners escaped from the police-station, and the Governor, himself unwell at the time, asked Fenton to give chase. He eventually overtook them in Cabeza del Mar, well inside Chile, and, when he approached them alone with a revolver, they gave themselves up. He also organized hunts and horseraces among the townsfolk, and played the piano, violin and flute. In 1900 he purchased 36 leagues of land at Jayak Aike, 125 kilometres west of Río Gallegos, eventually selling out to the Spaniards Rodolfo Suárez and Mauricio Bráun to buy an estancia in Tucumán in the north of Argentina, where he died in 1938.

One of Victor's nephews, Arthur Fenton, was seen less at Hill Station but was remembered as quite a character. He was born at Fenton Station in 1885 and at the age of twenty-five moved to Lago

Argentino and became the first resident of the estancia La Jeronima. He lived the most austere life, bathing daily in the cold waters of the Brazo Rico at dawn. He had very long hair, worn in red curls and tied back with a headband. He was a fantastic horseman, earning his keep horsebreeding and making tack, and was one of the few Patagonian pioneers never to keep a single sheep. Once a year he travelled to Río Gallegos with a troop of horses he had bred himself, and returned with provisions, most of which were eaten by the many visitors to his tents. By 1917 Lago Argentino was becoming too developed for his liking and the extermination of the pumas annoyed him. He sold out to the Stipicic brothers and moved north to the remoter Lago Cardiel, where he continued his horsebreeding. With a reputation for helping anyone in trouble, he died in Punta Arenas in 1937.

Percival Masters, born in Hampshire in 1876, arrived in Río Gallegos with his wife Jessie Elisabeth and, after working for a time at Cabo de los Virgenes, in 1904 moved to Lago Argentino by cart. He settled near Lago Roca, just south of the main lake, but left because of the devastation caused by the pumas among his sheep. He then settled at Cerro Frias, then at Calafate, and finally on the north bank of Lago Argentino, a five-hour sail from Punta Bandera. Once a year he visited Río Gallegos with his family, Percival, Herbert and Nelly, and, on one such visit, talked to Mabel Halliday about life in Lago Argentino: "Visits are rare: and time slips by with no alternative in an incredible solitude in this world of rapid communications."

Angus Martin, born on the Isle of Lewis in 1871, arrived in Patagonia in 1895 and worked as an employee for the Germans Curtze and Wahlen, to establish Las Horquetas, south of Lago Argentino. By 1907 he had saved enough to buy Chali Aike, between la Vanguardia and La Esperanza on the Río Coyle. He later told the Hallidays how for two years he lived in a tiny shelter made of six sheets of zinc, enduring temperatures of –20°C, his only supplies being from the Bráun y Blanchard store in Río Gallegos: "Beans, flour, hard tack and maté."

John Scott was probably the only other Patagonian pioneer from Dumfries. Born in 1863 near Loch Ettrick, Closeburn, he had moved to the Falklands in 1882, yet another shepherd under contract with FIC. But after seven years he became "tired of the isolated life and was forced to the conclusion that it would be nearly impossible to make a living there in any other way than as a shepherd". After taking 600 sheep to Herman Eberhart at Chymen

Aike in 1889, he bought 300 of them at 18 shillings a head and stayed there on condition that both men receive the same percentage per head of wool and enjoy the same increase of stock. The business arrangement did not work out. In 1894 he moved up the Río Gallegos in search of new land, with a bullock-cart and a few mares he had trained himself, with enough materials to build a small house and with provisions for a year. He settled at Bella Vista, some 70 kilometres from Río Gallegos, but, after losing 16,000 sheep, horses and cattle in the bad winter of 1902, he decided to sell out to the Bráun family and drove an arreo of his remaining 3,000 sheep to Los Machos near San Julián, the journey taking seven weeks. He arrived there in April when winter was approaching and just had time to put a roof on one room of an abode shack, and then settled down for the remaining three months of winter, with no fuel and very little food. At one stage he had to wade through deep snow to rescue his flocks, which had scattered for shelter to Salinas Grandes, a distance of over two leagues. Eventually his estancia prospered, and he became a much-respected member of Santa Cruz society, being on the jury of the first show held by the Sociedad Rural of Río Gallegos. He died in Dumfries in 1948.

Other visitors were the Tweedies from the Río Turbio, a tributary of the Río Gallegos, some 150 kilometres due west of Hill Station. John Tweedie, a Scot from Berwick-on-Tweed, had ridden south from Bahía Blanca in the late 1880s. He first took land at the north end of Lago Toro, but at the end of the century he bought the estancia Stag River at Río Turbio, his neighbour at Rospentek being the German Meyer who had introduced hares to Patagonia. An entry in Jamieson's diary for 20 February 1894 reads: "Tweedie left us for cordillera with sheep."

Other visitors were the Watson brothers, Charles and Roy. Charles, born in Southport, Lancashire in 1874, had joined the Merchant Navy and in 1891, as a cadet on the *Knight Commander* bound for San Francisco, jumped ship at Port Stanley. There he met and was taken under the wing of Frank Lewis on Keppel Island. In 1899 he moved to Porvenir on Tierra del Fuego and in 1902 he and his brother founded the estancia Rincón Grande on the banks of the Río Santa Cruz, some 130 kilometres upstream from Puerto Santa Cruz. His younger brother Roy, born in Freshfield, Lancashire in 1880, arrived in 1900 in Punta Arenas, where he met Robert Patterson and worked for a time at Mata Grande, near San Julián. He then became agent of a Liverpool company to buy wool in Patagonia, and made two annual trips between Punta Arenas and Puerto Deseado on horseback before finally joining his brother at Rincón Grande.

William Payne, born in 1885 in Leckhamstead, Berkshire, in 1896, joined his parents under contract with Arthur Waldron at Condor. In 1905 he established his own estancia Lago Roca, on the Brazo Rico of Lago Argentino, and married Gertrude Catalina Victoria Talbot, by whom he had twelve children. He was joined by John Atkinson, to whom in 1947 he sold his section of Lago Roca and settled in Aguas Vivas in the Río Pinturas in north-west Santa Cruz.

Another shepherd to begin work under contract with Waldron & Wood at Condor, having already been a shepherd with the Waldrons back in Berkshire, was George Drew, born in 1871 in Aldbourne, Wiltshire. In 1903 he terminated his contract and, in association with William Dickie, established Bon Accord at the eastern end of Lago Argentino. In 1928 he suffered a stroke and the farm was looked after by his eldest son, Charles.

Gerald John "Jack" Lively was remembered chiefly as the first settler of Lago San Martín. In 1893, at the age of fifteen, Jack was an apprentice on an old windjammer, the *Sir George F. Seymour*. Passing through the Straits, he was so taken with what he saw that he decided to settle there. He worked for a few weeks for José Menéndez at San Gregorio, but, "owing to a slight difference of opinion with the cook", he left and rashly set out to explore southern Patagonia on foot. Fortunately he was picked up by an old Falkland gaucho, Dick Pitaluga, who lent him a horse and took him on the first drive of sheep to populate Ultima Esperanza at the west end of the Straits. In 1895 he worked for a while with Ernest Cattle at Lago Argentino, driving the first-ever bullock-wagons from there to Río Gallegos. The following year he joined Fred Otten, a taxidermist sent from Hamburg by Hagenberg & Co. to collect animals, and explored the cordillera as far as Lago San Martín. In 1897 he again worked for Cattle in Lago Argentino and with his three brothers, William, Joseph and Robert, whom he had persuaded to join him, drove an arreo of mares from Fenton Station in the Straits near Punta Arenas to Tres Pasos at Ultima Esperanza. From 1899 to 1901 he fought in the Boer War in the same regiment as Winston Churchill and claimed that, when Churchill was wounded at Vaalkrantz in 1900, he was among those who helped him back to the stretcher-bearers. In 1901 he set up an estancia at Tres Pasos with his brothers, but after several bad years and after working with the Boundary Commission, he moved to inspect Lago San Martín with his brother Percy, six horses, a mule and four hunting-dogs. They managed to hunt successfully: "There was no need for bullet or bolas; one dog chased, the others cut it off." They collected ostrich-feathers, which they eventually exchanged for a few sheep in Río Gallegos, drove them back and established the first sheepfarm at Lago San Martín. After such a life, Jack was naturally scathing of what he considered amateur sportsmen: "The sportsmen of little England, who cannot hunt rabbits for more than two hours without delving into a picnic-basket full of titbits, without a steward following them with an urn of tea."

Another visitor was Lionel Harris. Born in Aldershot, Hampshire, he trained as an accountant but soon became bored and in 1907 arrived in Punta Arenas. He settled for a while with Hugo Lively at Paso Piedrabuena, and then set out for Lago San Martín, where he worked as peon and shepherd at Lago Tar, a small lake east of Lago San Martín. He returned to England to fight in the First World War, during which he was gassed, and so returned to Patagonia for an outdoor life. He managed several estancias, was British vice-consul in Puerto Santa Cruz from 1922 to 1948, and finally settled in a military zone near Comodoro Rivadavia.

James Lewis and his wife, Eleanor Nell Britton, and their son William sailed in 1871 from Bristol to join Rev. Thomas Bridges, father of Lucas and Despard, then in charge of the Patagonian Missionary headquarters on Tierra del Fuego. Frank Ushuaia was born in the Falklands because Tierra del Fuego itself had no maternity facilities. In 1890 he became representative of Alex Robertson of Oban for selling dip and as a result covered the area from Río Gallegos to the Straits. In 1894 he purchased 1,500 sheep at Cabeza del Mar in Chile and, with his brother, drove them north to the Río Santa Cruz. They had to cross the river in midwinter, so they constructed a type of wherry with help from the Tehuelches. They finally settled at Cañadón El Toro, some 25 kilometres west of Puerto Santa Cruz, in partnership with Wickham Bertrand. In 1905 the business association was dissolved, William remaining at Cañadón El Toro, Frank going to Corpen Aike, later renamed La Margarita, east of Lago Viedma.

Herbert J. Elbourne was born in London in 1879, but in 1903, because of ill-health, he sought a dry, cold climate and so fled to Patagonia. He was on board the *California*, which ran aground in Bahía Blanca, and was saved by the *Ushuaia*, in which he continued to Puerto Madryn and worked for the Welsh Ferrocaril Central del Chubut, where he married Alice Maud Pickering, from an important family of settlers. In 1909 he joined the Sociedad Anónima Importadora y Exportadora de la Patagonia, and became their accountant in Puerto Santa Cruz and Río Gallegos. In 1920, with others, he founded an estancia near Lago Belgrano.

Others who came from the Falkland Islands (and the area in which they finally settled): William Douglas (Morros Grandes, near Río Gallegos); William Hubbard (Río Chico); Wickham and Roy Bertrand, and William Betts (Río Santa Cruz); John Blakeley, Charles Hansen, Conrad Rowlands and William Watson (San Julián).

Others who came directly from Scotland: William Ness (Río Gallegos); John MacCormack, Robert Macdonald and John MacCleod (Río Coyle); Angus and Hugh Macpherson (Ultima Esperanza); Donald and Alexander MacBean, and John Oman (Río Chico); William and Donald Bain, John MacKae and Robert Nicholson (Puerto Deseado); George Anderson, Alexander Finlayson, John Frazer, William and James Hope, Andrew Kyle, James MacKay, John MacLean, Donald Munro, William Reid, William Wallis and Simon Wilson (San Julián).

Others who came directly from England: Philip Marshall and James Perks (Río Coyle); Alfred Barclay, Robert Blake, Edwin King, John Merrick, Patrick Orr, Charles Veley and Cecil Withers (San Julián); John Atkinson, George Gregory, Walter Reeve and Claud Waring (Lago Argentino); George and Walter Harris (Río Turbio); Herbert Bostock, Ernest Hartlock, Albert Mason, William Wallace and George Wilbey (Lago Tar); James Carpenter, Alfred Giles and Percy, Hugh and Norman Naish (Ultima Esperanza); T. C. Burberry, Thomas Foreman, John Harvey, James Hudson and George Woolven (Río Santa Cruz); William Ford, Harry Park, Oliver Sherman and Arthur Wing (Lago San Martin).

By no means all of the settlers were of British origin: from Ireland, John O'Keefe (Río Viedma); from Australia, Hugh Denniston (Río Santa Cruz) and August Moy (Río Chico); from Newfoundland, William Squires (Río Santa Ana) and from India, William Lippert (San Julián).

Although most of the Hallidays' friends and acquaintances were of British stock, there were plenty of settlers from other European countries, especially from Germany. Enrique Bitsch arrived in Patagonia in 1887 at the age of 29. After travelling from Punta Arenas to Puerto Santa Cruz with one tent and virtually no food he finally bought sheep from Herman Eberhart and formed an estancia near Bella Vista with his brother, returning home for several years to Friedrichstadt in Schleswig-Hölstein before re-settling in Patagonia because of difficulties in Germany. Walter Curtze arrived in Punta Arenas in 1885 as shipping agent for the Kosmos Line of Hamburg. In 1893 he married Mary Julie Williams from the Falkland Islands and founded the estancia Monte León, near the Atlantic coast some ten kilometres south of Puerto Santa Cruz. He also helped his father-in-law Charles Williams establish himself at Kilik Aike Sud in succession to Rodolfo Suárez and then returned to Punta Arenas, where he founded an electricity company and the Banco de Punta Arenas, and was president of the German school and the German club till his death in 1920. Carlos Führ, born in Germany in 1861, was an early settler near Güer Aike, then at Cerro Castillo, just across the frontier in Chile near Lago del Toro. In 1904 he moved to La Leona at the far east of Lago Argentino. He built a hotel and ferry, and preferred to call the place Carlos Führ. He had a reputation among other settlers for being lazy and boastful, claiming that he had once caught the infamous Ascencio Brunel red-handed. In 1912 he sold out and returned to Germany, but eventually came back to Río Gallegos, where he died "in abject poverty". Rudolph Hamann, born in Schleswig-Hölstein, at first worked on the railways in Rosario, central Argentina. In 1887 he arrived in Río Gallegos aboard the *Villarino* and met Lieut. Castillo, who was sent by the Argentine government to explore the Andes of the Río Gallegos which resulted in the discovery of coal in the Río Turbio. Hamann joined the expedition and they returned by descending the Río Gallegos as far as Güer Aike in a canoe dug out of a tree-trunk. In 1890 he established a sawmill in Punta Arenas and finally settled at Laguna de Oro, in mid-camp halfway between Puerto Coyle and Lago Argentino. Ernest von Heinz, a cousin of Herman Eberhart, arrived in Río Gallegos and accompanied Eberhart on his move to Ultima Esperanza, where he discovered the famous "cave of the mylodon". He finally settled in Cerro Palenque, east of Lago Argentino, with a fellow-German, Rodolfo Stubenrauch. Augusto

Kark was one of the few settlers to arrive in Patagonia with considerable capital and he at once purchased Markatch Aike on the Río Chico south of Río Gallegos. He married one of Eberhart's daughters and formed a company with his compatriots Bernardo Osenbrüg and Pablo Lenzer, who had come south after their crops in Córdoba were devoured by locusts.

There were also several Yugoslavs, such as the Stipicic brothers, José and Jerónimo, who arrived in Patagonia in 1899. After working for a short time at the estancia Condor, they searched for gold in Cabo de los Virgenes and then ran the slaughter-house in Punta Arenas. In 1913 they acquired the rights of Ernest Cattle in Cerro Buenos Aires, later those of Arthur Fenton at La Jerónima and of a Frenchman, Lesseur, at Alta Vista. In 1926 their property was severely reduced under the Ley de Tierras, restricting the amount of land any individual might own. They remained, however, among the wealthiest of all Patagonian pioneers. There were a handful of Scandinavians. Like the poet Madsen, Rodolfo Mortensen was born in Jutland, and accompanied a German, Carlos Siewert, who was commissioned to survey the 400 square leagues conceded to military leaders under the Grümbein law. In 1907 he purchased camp 150 kilometres upriver from Puerto Santa Cruz. Halvor Halvorsen was a Norwegian who in 1916 populated the south shores of Lago Viedma, his first shelter being a calafate bush covered with mareskin in which Ascencio Brunel had reputedly once lived. Cayetano Cornet d'Hunval was a Frenchman who, after living for several years in central Argentina, arrived in Patagonia in 1910, settling at Laguna Benito, where there were still plenty of Tehuelche tents, and bringing all his provisions in a dray from Puerto Santa Cruz. Lorenzo Toso was an Italian who arrived at Lago Argentino in 1908 after being manager at the Douglas's estancia Esperanza. He built a small boat, sailed far up the lake and founded María Antonia, later called La Argentina, on the north bank.

There were plenty of Spaniards. Iván Noya, a surviving member of the official colony of Puerto Deseado in 1885, was in 1889 contracted by Governor Lista to build a quay at Río Gallegos and a year later opened the first important business-house in the town. In 1885 he founded the estancia Paso del Medio on the south bank of the Río Gallegos about 50 kilometres from the town. He took an active interest in politics, was president of the Town Council and Sociedad Rural, and during the labour troubles of 1921–2 was president of La Liga Patriótica and was said to have lavished his own money on troops sent from Buenos Aires to quell the strikes. Rodolfo Suárez, born in Asturias in 1866, arrived in Punta Arenas in 1882 and spent five years at Cabo Virgenes looking for gold and then worked at the estancia Tres Chorillos on the Straits. In 1890 he set up a small business house in Río Gallegos and then established himself at Kilik Aike Sud with 500 sheep imported from the Falklands. In 1895 he moved to Las Horquetas, 80 kilometres west of Río Gallegos, with 2,000 merinos, 200 cows and 200 horses bought from George McGeorge, his flocks eventually expanding to 42,000, with Romney Marsh and Corriedale the most popular breeds for crossing.

Bibliography

AKERS, C.E., *Argentine, Patagonian and Chilian Sketches*. London, 1893.

ANSON, (George) Lord, *A Voyage to the South-Seas and to many other parts of the World, performed in 1740-44*. London, 1744.

ARGENTINA AUSTRAL. Monthly periodical published by La Sociedad Anónima Importadora y Exportadora de la Patagonia, Buenos Aires, July 1929 (et seq).

BAYER, Osvaldo, *Los Vengadores de la Patagonia Trágica*. 2 vols, Buenos Aires, 1972.

BEERBOHM, Julius, *Wanderings in Patagonia or Life Among the Ostrich-Hunters*. London, 1879.

BORRERO, José María, *La Patagonia Trágica*. Buenos Aires, 1928.

BOURNE, Benjamin Franklin, *The Giants of Patagonia. Captain Bourne's Account of his Captivity amongst the Extraordinary Savages of Patagonia*. London, 1853.

BOWEN, E.G., *The Welsh Colony in Patagonia 1865-85: a study in Historical Geography. Geographical Journal*, London, March 1966.

BOYSON, V.F., *The Falkland Islands*. Oxford, 1924.

BRÁUN MENÉNDEZ, Armando, *Pequeña Historia Magallánica*. Buenos Aires, 1945.

BRIDGES, E. Lucas, *Uttermost Part of the Earth*. London, 1948.

---, two short chapters omitted from *Uttermost Part of the Earth* (from a copy in possession of Helen Sinclair de Halliday).

BULKELEY, John, & CUMMINS, John, *A Voyage to the South Seas*. London, 1743.

BYRON, John: see *Byron's Journal of His Circumnavigation 1764-1766*, ed. Robert E. Gallagher. Cambridge (Hakluyt Society) 1964.

CATLIN, George, *Last Rambles amongst the Indians of the Rocky Mountains and the Andes*, London, 1868.

CAVENDISH, Thomas: see *Voyages of the Elizabethan Seamen to America*, ed. E.J. Payne. Oxford, 1900.

CAWKELL, M.B.R., *The Falkland Islands*. London, 1960.

CHAMBERS, R. (ed.), *The Book of Days. A Miscellany of Popular Antiquities*. London & Edinburgh, 1888.

CHATWIN, Bruce, *In Patagonia*. London, 1977.

CHIDLEY, John: see *A Brief Relation of a Voyage of the Delight...begun in the year 1589* written by W. Magoths. *Hakluyt's Voyages* (vol. iii), Everyman, London, 1907.

COAN, Titus, *Adventures in Patagonia. A Missionary's Exploring Trip*. New York, 1880.

COPPINGER, Robert O., *Notes on the Natural History of the Strait of Magellan and West Coast of Patagonia*. Edinburgh, 1871.

DARWIN, Charles, *Journals and Remarks, 1832-6*; vol. III of the *Narrative of the Surveying Voyages of the Adventure and Beagle, on the Shores of South America*. London, 1839.

– – –, *The Life and Letters of Charles Darwin*, ed. Francis Darwin. London, 1888.

DIXIE, Lady Florence, *Across Patagonia*. London, 1880.

– – –, *Aniwee; or, the Warrior Queen. A tale of the Araucanian Indians and the mythical Trauco people*. London, 1890.

– – –, *The Young Castaways, or, the Child Hunters of Patagonia*. London, 1890.

FALCÓN, Edelmiro A. Correa, *La Patagonia Argentina*. Buenos Aires, 1924.

– – –, *Vidas Patagónicas*. Buenos Aires, 1950.

FALKNER, Thomas, *A Description of Patagonia, and the Adjoining Parts of South America*. Hereford, 1774.

FITZROY, Robert, *Proceedings of the Beagle's Second Voyage to South America 1831-6*. London, 1839.

FLETCHER, Francis, *The World Encompassed by Sir Francis Drake*. London, 1628.

FOREIGN AND COMMONWEALTH OFFICE, *Living in the Falkland Islands*. Revised ed. London, 1973.

GOEBEL, Julius, *The Struggle for the Falkland Islands*. New Haven, 1927.

HADFIELD, William, *Brazil, the River Plate, and the Falkland Islands*. London, 1854.

HALLIDAY, Mabel, diaries, notes and letters by, to and from members of her family. 1890 et seq.

HAWKINS, Sir Richard, *Observations in his Voyage in the South Sea. A.D.1593*. London, 1622.

HERNÁNDEZ, José, *El Gaucho Martín Fierro*, tr. Walter Owen. Bilingual ed., Buenos Aires, 1967.

HOLDICH, Thomas, *The Countries of the King's Award*. London, 1904.

HUDSON, W.H., *Idle Days in Patagonia*. London, 1893.

JAMIESON, Henry, *Diary*. 1892 et seq.

JOHNSON, Samuel, *Thoughts on the Late Transactions respecting the Falkland Islands*. London, 1771.

JONES, Tom P., *Patagonian Panorama*. Bournemouth, 1961.

MADSEN, Andreas, *Patagonian Poems*. Lago Viedma, 1943.

MAGELLAN, Ferdinand: see *Magellan's Voyage Round the World*, ed. Charles E. Nowell. Evanston, 1962.

MANSFELD, Franz, *El Bosque Petrificado de Santa Cruz*. Buenos Aires, 1940.

MATTHEWS, A. (Rev.), *Crónica de la Colonia Galesa de la Patagonia*. tr. F.E. Roberts. Buenos Aires, 1954.

MORALES, Ernesto, *La Ciudad Encantada de la Patagonia*. Buenos Aires, 1944.

MORENO, Francisco, *Viaje a la Patagonia Austral*. Buenos Aires, 1879.

MOYANO, Carlos, *Exploración de los Rios Gallegos, Coile, Santa Cruz y Canales del Pacífico*. Buenos Aires, 1887.

MULHALL, M.G. & E.T., *Handbook of the River Plate*. Buenos Aires, 1863 (et seq.).

MUSTERS, George Chaworth, *At Home with the Patagonians*. London, 1871.

NARBOROUGH, Sir John, *An Account of Several Late Voyages and Discoveries to the South and North*. London, 1694.

NERUDA, Pablo, *Selected Poems*, ed. Nathaniel Tarn. London, 1970.

NEW LEAVES. The Aberdour House School Magazine. Dumfries, 1903 (et seq.).

NEWTON, Alfred, *A Dictionary of Birds*. London, 1896.

ORLLIE-ANTOINE: see *Pequeña Historia Patagonica* by Armando Bráun Menéndez. Buenos Aires, 1945.

– – –, *El Ingenioso Hidalgo don Orllie I* by Thomás Eloy Martínez. Siete Dias. Dec. 1971, Buenos Aires.

PRICHARD, H. Hesketh, *Through the Heart of Patagonia*. London, 1902.

SCHMID, Teofilo: see *The Voice of Pity for South America*. Published by the Patagonian Missionary Society. London, 1858 (et seq.).

SCHOUTEN, William, *The Relation of a Wonderful Voyage*. London, 1619.

SCHULZ, Gustav, *Falkland Islands. South America*. Printed Leipzig, published London c. 1880.

The STANDARD, and the BUENOS AIRES HERALD. English-language newspapers of Buenos Aires.

TSCHIFFELY, A.F., *This Way Southward*, London 1945.

– – –, *Tschiffely's Ride*. London, 1952.

ULLOA, C. Juan & A. de, *A Voyage to South America*, tr. J. Adams. London, 1758.

WILLIAMS, Sir Ralph, *How I Became a Governor*. London, 1913.

WOOD, Edward J., *Giants and Dwarfs*. London, 1868.

CHILI
LA PLATA
P. Argentino
Chiloe
Villarica V.
Valdivia
R. Valdivia
S. Pedro B.
R. Bueno
Pt. Estaquilla
R. Osorno
L. Nahuelhuapi
Pt. Godoy
L. Purailla
Pt. Guabun
S. Carlos
C. Matalqui
C. Ypuntia
CHILOE
Comau Inl.
Reniluc Inl.
V. Minchinmadom
Corcobado Vol.
M. Yanteles
Palena B.
Huafo or
No Man's I.
Guaytecas Is.
M. Melinoyu
Refugio B.
M. Menlolat
M. Maca
R. Aysen
Guaiad
Narborough I.
Socorro I.
Nimula Ch.
Darwin Ch.
Garrido
Anna Pink B.
Skering L.
Hellver B.
Peninsula
of Tres Montes
S. Andres
C. Raper
C. Tres Montes
S. Esteven
Xavier I.
Jesuit Sd.
Port Rey
R. Sinfondo
L. S. Rafael
Aldunate Inlet
G. OF PENAS
Guaineco I.
P. S. Barbara
Baker I.
Campana I.
Paralle
Dyneley
C. Montague
Picton Opening
M. Corso
G. of Trinidad
Archipelago
of
Madre de Dios
G. Santiago
Concepcion Str.
HANOVER I.
S. Blas Ch.
Cambridge I.
Nelson Str.
Diana Pk.
C. Victory
Narborough I.
STR. OF MAGALHAEN'S
C. Pillar
C. Descalio
Otway Bay
Rice Trevor I.
C. Tate
Breaker R.
C. Gloucester
Gratton I.
Stokes Bay
Noir I.
Kemper I.
Furry
Cockburn Chan.
Desolation I.
Adventure Passage
Londonderry I.
Sandwich I.
York Minster
Darwin Sound
Christmas
Christmas Sound
WELLINGTON I.
Messier Channel
Fallos Channel
Iceberg Sd.
Lion B.
Eyre Sd.
Falcon Inl.
Castle Hill
Canning I.
San Andres Sd.
Pitt Chan.
Peel Inl.
M. Stokes
L. Capar
R. Chalia
R. Sta. Cruz
PATAGONIA
MOLUCHE INDIANS
Chullilon Indians
Puelche or Eastern People
Puelches of Indians or wandering Tribes of
Inhabited by wandering Tribes of Indians
Tehuelhuet or Southern People
P. S. Antonio
GULF OF SAN MATIAS
M. S. Antonio
Sierra Pt.
Pt. Quiroga
Pt. Norte
P. S. Josef
Pt. Bajos
Pen. of S. Josef or Valdes
Pt. Delgada
Pt. Ninfas
Nuevo Gulf
B. Chupat
Prop'd. Settlem't.
Pt. Casino
Engaño R.
Jaibos Hd.
M. Triste
Pt. Atlas
R. Camerones
C. Raso
Vera B.
Olmos
Port S. Elena
P. Malaspina
Camerones B.
Quintanar
C. Two Bays
Garden I.
Leones I.
Salamanca P.S.
C. Cristzabal
Menkuares
Ali Rocks
Pt. Marques
GULF OF St. GEORGE
Pt. Murphy
Langara B.
Plaza Hd.
Pt. Casamayor
Nava Hd.
C. Three Points
Byron Sd.
C. Blanco
Mizarredo B.
River Pd.
P. Desire R.
Port Desire
Spiring It.
Watchman C.
Desvelas B.
Bird I.
Pt. Lookout
Port S. Julian
Francisco de Paulo
Beachy Hd.
Port Sta. Cruz
Lion M.
Observation M.
C. Redonds
Coy Inlet
Tiger M.
C. Fairweather
Port Gallegos
Friars
M. Aymond
Possession B.
Obstruction Sd.
C. Virgins
STRAIT OF MAGALHAEN'S
Orange
C. Espiritu Santo
Pt. Arcina
Sebastian B.
EASTERN
OR KING CHARLES SOUTH LAND
TIERRA DEL FUEGO
C. Sunday
C. Peñas
C. Medio
Brunswick P.
Useless B.
Dawson I.
Lawrence I.
C. S. Diego
Le Maire Strait
C. S. Anthony
C. New Yorks I.
St. John C. & Har.
C. Bartholomew
Cape Vancouver
Staten I.
Adventure Passage
Gordon I.
HOSTE I.
Navarin I.
Picton I.
New I.
Lennox I.
Evout I.
Nassau B.
Wollaston I.
Franklin St.
Cape Horn
TIERRA DE L FUEGO
SCALE
50 Miles
Beagle Chan.
40
75
65
55
65
Deer
St. Blas B.
Carmen
Pt. Rasa
Sisters
Main Pt.
Belen Bluff
Tide Ch.
Fort
Oacen B.
Falso B.
Brightman Inl.
Union B.
Incgada B.